MR. RHEE'S BRILLIANT
MATH SERIES

MW01565603

Solomon Academy

7 FULL-LENGTH

Practice Tests for the AP Calculus BC Exam Multiple Choice Sections

By Brian Rhee

Reflects Current AP Calculus BC Exam

7 Practice Tests Contain the Most Up-To-Date Types of Problems

Provides Only Answer Keys

Legal Notice

Published by: Solomon Academy
First Edition
ISBN-13: 978-1977881069
ISBN-10: 1977881068

About Author

Brian(Yeon) Rhee obtained a Masters of Arts Degree in Statistics at Columbia University, NY. He served as the Mathematical Statistician at the Bureau of Labor Statistics, DC. He is the Head Academic Director at Solomon Academy due to his devotion to the community coupled with his passion for teaching. His mission is to help students of all confidence level excel in academia to build a strong foundation in character, knowledge, and wisdom. Now, Solomon academy is known as the best academy specialized in Math in Northern Virginia.

Brian Rhee has published more than ten books. The titles of his books are 7 full-length practice tests for the AP Calculus AB/BC Multiple choice sections, AP Calculus, SAT 1 Math, SAT 2 Math level 2, 12 full-length practice tests for the SAT 2 Math Level 2, SHSAT/TJHSST Math workbook, and IAAT (Iowa Algebra Aptitude Test) Volume 1 and 2, CogAT form 7 Level 8, and NNAT 2 Level B Grade 1. He's currently working on other math books which will be introduced in the near future.

Brian Rhee has more than twenty years of teaching experience in math. He has been one of the most popular tutors among TJHSST (Thomas Jefferson High School For Science and Technology) students. Currently, he is developing many online math courses with www.masterprep.net for AP Calculus AB and BC, SAT 2 Math level 2 test, and other various math subjects.

Acknowledgements

I wish to acknowledge my deepest appreciation to my wife, Sookyung, who has continuously given me wholehearted support, encouragement, and love. Without you, I could not have completed this book.

Thank you to my sons, Joshua and Jason, who have given me big smiles and inspiration. I love you all.

ABOUT AP Calculus AB and BC Exams

The AP Calculus AB and BC exams are intended to measure the extent to which a student has mastered the subject matters of the AP Calculus course. Although the AP Calculus courses focus on differential and integral calculus, students need strong foundations in Algebra, Geometry, and Trigonometry.

Each AP exam is 3 hours and 15 minutes long, and its format is as follows:

Section 1 Multiple choice — 45 Questions (1 hour and 45 minutes)

Part A: 30 questions for 60 minutes (calculator is not permitted)

Part B: 15 questions for 45 minutes (graphing calculator is required)

Section 2 Free Response — 6 Questions (1 hour and 30 minutes)

Part A: 2 questions for 30 minutes (graphing calculator is required)

Part B: 4 questions for 60 minutes (calculator is not permitted)

Contents

CALCULUS BC TEST 1
SECTION I, Part A
Time — 60 minutes
Number of questions — 30

A CALCULATOR MAY NOT BE USED ON THIS PART OF THE EXAM.

Directions: Solve each of the following problems using the available space for scratch work. Choose the best answer among the answer choices given and fill in the corresponding circle on the answer sheet.

1. $\int 2\sqrt{x}(x^2 - x)dx =$

(A) $\frac{4}{7}x^{7/2} - \frac{4}{5}x^{5/2} + C$

(B) $\frac{1}{7}x^{7/2} - \frac{1}{5}x^{5/2} + C$

(C) $2x^{5/2} - \frac{2}{3}x^{3/2} + C$

(D) $\frac{2}{3}x^{3/2} - \frac{3}{4}x^{1/2} + C$

2. If $y = \dfrac{2}{\sqrt{x}} - \dfrac{4}{\sqrt[3]{x}}$, then $\dfrac{dy}{dx} =$

(A) $4\sqrt{x} - 6x^{2/3}$

(B) $-\dfrac{\sqrt{x}}{2} + \dfrac{x^{3/2}}{4}$

(C) $-\dfrac{1}{x\sqrt{x}} + \dfrac{4}{3x\sqrt[3]{x}}$

(D) $x^{3/2} + \dfrac{4}{3}x^{-4/3}$

---

3. Which of the following is the equation of the tangent line to the curve $\dfrac{1}{y} + xy = 3$ at the point $(2, 1)$?

(A) $y - 1 = -(x - 2)$

(B) $y - 1 = -\dfrac{1}{3}(x - 2)$

(C) $y - 1 = \dfrac{1}{3}(x - 2)$

(D) $y - 1 = (x - 2)$

4. If $f(x) = \dfrac{\sin 2x}{2\sin x}$, then $f'(x) =$

(A) $-\sin x$

(B) $-\tan x$

(C) $\csc^2 x$

(D) $\sec^2 x$

---

5. Suppose a population P grows according to the differential equation $\dfrac{dP}{dt} = \dfrac{1}{4}P - \dfrac{P^2}{1000}$. The growth rate of the population is the fastest when the population reaches which of the following?

(A) $P = 100$

(B) $P = 125$

(C) $P = 250$

(D) $P = 1000$

6. Which of the following is the value of $\sum_{n=1}^{\infty}(-1)^n \frac{3^n}{4^{n+1}}$?

(A) $\frac{5}{7}$ (B) $\frac{5}{28}$ (C) $-\frac{3}{7}$ (D) $-\frac{3}{28}$

---

7. Let $y = f(x)$ be the solution to the differential equation $\frac{dy}{dx} = x + y$ with initial condition $f(3) = -1$. Which of the following is the approximation of $f(4)$ using the Euler's method with two steps of equal length?

(A) 2.5 (B) 2.25 (C) 2 (D) 1.75

8. If $\int_0^b \frac{x^2}{x^3+1}\,dx = \frac{1}{3}\ln 9$, where $b > 0$, then $b =$

(A) 1 (B) $\frac{3}{2}$ (C) 2 (D) $\frac{5}{2}$

$$f(x) = \begin{cases} 3, & 0 \le x < 2 \\ x, & x \ge 2 \end{cases}$$

9. If the function f is defined above, which of the following is the value of $\int_0^4 f(x)\,dx$?

(A) 9 (B) 12 (C) 15 (D) nonexistent

10. If a particle is moving along the curve defined by the parametric equations $x(t) = 1 + \cos t$ and $y(t) = 2 + \sin^2 t$, which of the following best represents the total distance that the particle traveled from $t = 1$ to $t = 2$?

(A) $\int_1^2 \sqrt{(1+\cos t)^2 + (2+\sin^2 t)^2}\, dt$

(B) $\int_1^2 \sqrt{\cos^2 t + \sin^2 t}\, dt$

(C) $\int_1^2 \sqrt{\cos^2 t + 4\sin^2 t}\, dt$

(D) $\int_1^2 \sqrt{\sin^2 t + \sin^2 2t}\, dt$

---

11. Which of the following does $\lim_{x\to\infty} f(x) = \infty$?

I. $f(x) = \dfrac{2^x}{x^2}$ II. $f(x) = \dfrac{\ln x}{e^x}$ II. $f(x) = \left(1 + \dfrac{1}{x}\right)^x$

(A) I only

(B) III only

(C) I and II only

(D) II and III only

12. Which of the following are all values of x for which $\sum_{n=1}^{\infty} \frac{(-1)^n(x+2)^n}{4^{n+1}}$ converges?

(A) $-6 < x < 2$

(B) $-6 \leq x < 2$

(C) $-2 < x < 6$

(D) $-2 < x \leq 6$

---

13. If $a > 0$ and $\lim_{x\to\infty} \frac{\int_1^x 1+t^2\, dt}{ax^3} = \frac{1}{6}$, then which of the following is the value of a ?

(A) 6

(B) 5

(C) 4

(D) 2

$$e^x = 1 + x + \frac{x^2}{2} + \frac{x^3}{6} + \cdots$$

14. The Taylor series for e^x about $x = 0$ is shown above. Which of the following is the Taylor series for e^{-3x} about $x = 0$?

(A) $-3 - 3x - \frac{3x^2}{2} - \frac{3x^3}{6} + \cdots$

(B) $1 - 3x + \frac{9x^2}{2} - \frac{27x^3}{6} + \cdots$

(C) $1 - \frac{x}{2} - \frac{9x^2}{6} - \frac{27x^3}{24} + \cdots$

(D) $3 + \frac{3x}{2} + \frac{3x^2}{6} + \frac{3x^3}{24} + \cdots$

---

15. If $\lim_{h \to 0} \frac{\arctan(a+h) - \arctan(a)}{h} = \frac{1}{4}$, then which of the following could be the value of a ?

(A) $\sqrt{3}$ (B) $\sqrt{2}$ (C) $\frac{\sqrt{3}}{2}$ (D) $\frac{\sqrt{2}}{2}$

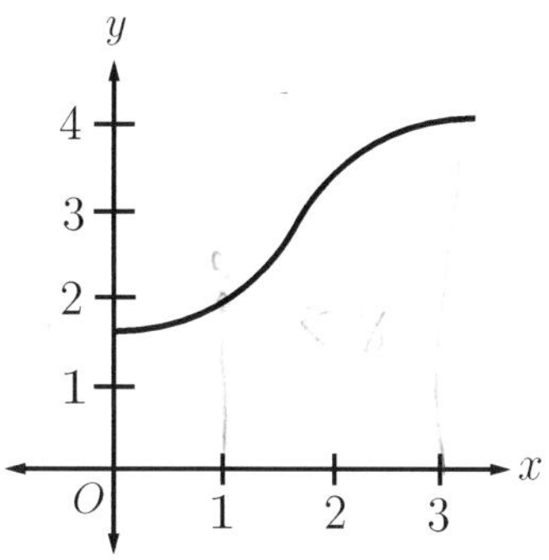

Graph of f'

16. The graph of f', the derivative of f, is shown in the figure above. If $f(1) = 5$, all of the following could be the value of $f(3)$ EXCEPT

(A) 9.1

(B) 11.8

(C) 12.5

(D) 13.2

---

17. The third-degree Taylor polynomial for a function f centered at 1 is $(x-1) - \frac{(x-1)^2}{2} + \frac{(x-1)^3}{3}$. Which of the following is the value of $f'''(1)$?

(A) 2

(B) 1

(C) $\frac{1}{2}$

(D) $\frac{1}{18}$

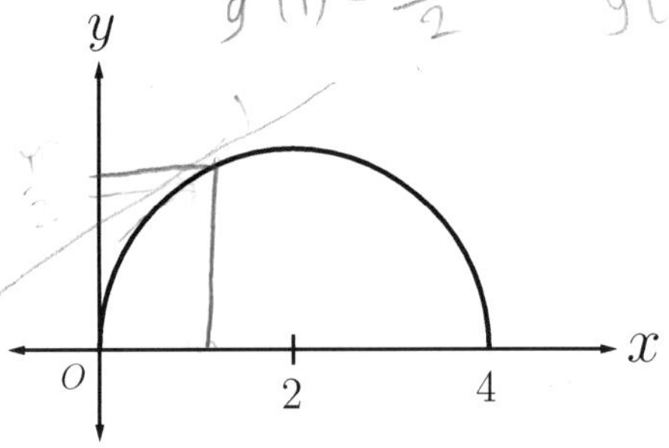

Graph of f

18. The graph of the function f shown above is a semi-circle of radius 2 with center at $(2, 0)$. If the function g is defined by $g(x) = \int_0^{2x} f(t)\,dt$, which of the following is an equation of the tangent line to the graph of g at $x = 1$?

(A) $y - \frac{\pi}{2} = -2(x - 1)$

(B) $y - \frac{\pi}{2} = x - 1$

(C) $y - \pi = 4(x - 1)$

(D) $y - \pi = 6(x - 1)$

19. For $x > 2$, which of the following is the antiderivative of $\frac{4}{4 - x^2}$?

(A) $2\tan^{-1}\left(\frac{x}{2}\right) + C$

(B) $4\ln\left(\frac{x+2}{x-2}\right) + C$

(C) $\ln(x + 2) - \ln(x - 2) + C$

(D) $\ln\left(\frac{x-2}{x+2}\right) + C$

x	f	g	f'	g'
3	2	5	4	-1
5	-3	1	-2	5

20. The table above shows values of f, g, f' and g' for selected values of x. If the function h is defined by $h(x) = f(g(x))$, which of the following is the value of $h'(3)$?

(A) 6 (B) 2 (C) 1 (D) -2

---

21. The length of the base of a triangle, b, is increasing at a rate of 2 cm/min and the height of the same triangle, h, is decreasing at a rate of 1 cm/min. Which of the following is the rate at which the area of the triangle, A, is decreasing when the area of the triangle is 20 cm^2/min and the length of the base is 10 cm ?

(A) $\dfrac{dA}{dt} = -\dfrac{5}{2}\ \text{cm}^2/\text{min}$

(B) $\dfrac{dA}{dt} = -2\ \text{cm}^2/\text{min}$

(C) $\dfrac{dA}{dt} = -\dfrac{3}{2}\ \text{cm}^2/\text{min}$

(D) $\dfrac{dA}{dt} = -1\ \text{cm}^2/\text{min}$

22. If the function f is defined by $f(x) = 3x^3 - 9x^2 - 27x + 5$, which of the following is the absolute maximum value of f on the closed interval $[-2, 2]$?

(A) -61 (B) -1 (C) 11 (D) 20

23. Which of the following is the coefficient of x^{10} in the Taylor series for $5 \sin x^2$ about $x = 0$?

(A) $\frac{6}{5}$ (B) $\frac{3}{8}$ (C) $\frac{1}{24}$ (D) $\frac{1}{600}$

24. Which of the following is the solution of the differential equation $\dfrac{dy}{dx} = \dfrac{x^2+1}{\sec^2 y}$ with the initial condition $y(3) = \dfrac{\pi}{4}$?

(A) $\tan y = \dfrac{1}{3}x^3 + x - 11$

(B) $\tan y = -\dfrac{1}{3}x^3 + x + 9$

(C) $y = \tan\left(\dfrac{1}{3}x^3 + x - 11\right)$

(D) $y = -\tan\left(\dfrac{1}{3}x^3 + x + 9\right)$

---

25. Which of the following value does the improper integral $\displaystyle\int_1^{\infty} xe^{-x^2}\,dx$ converge?

(A) $\dfrac{1}{e}$ (B) $\dfrac{1}{2e}$ (C) $\dfrac{1}{4e}$ (D) divergent

26. At which of the following point (x, y) is the tangent line to the curve $x^3 - 6xy + y^3 = 0$ vertical?

(A) $(2, -2)$ (B) $(2, -1)$ (C) $(2, 0)$ (D) $(2, 1)$

---

27. A region is enclosed by $y = \sin 2x$ and the x-axis for $0 \le x \le \dfrac{\pi}{2}$. Which of the following integral best represents the volume V obtained by rotating the region about the line $x = -1$?

(A) $V = \pi \displaystyle\int_{-1}^{\frac{\pi}{2}} (\sin 2y)^2 - 1^2\, dy$

(B) $V = \pi \displaystyle\int_{0}^{\frac{\pi}{2}} (\sin 2y + 1)^2 - 1^2\, dy$

(C) $V = 2\pi \displaystyle\int_{-1}^{\frac{\pi}{2}} (x + 1)(\sin 2x + 1)\, dx$

(D) $V = 2\pi \displaystyle\int_{0}^{\frac{\pi}{2}} (x + 1) \sin 2x\, dx$

28. Which of the following series are absolutely convergent?

I. $\sum_{n=1}^{\infty}(-1)^n \frac{1}{\sqrt{n}}$ II. $\sum_{n=1}^{\infty} \frac{\cos n\pi}{n^2}$ II. $\sum_{n=1}^{\infty} \frac{n^2+n}{2n^2+1}$

(A) I only

(B) II only

(C) I and III only

(D) II and III only

29. If $x(t) = \frac{1}{2}t^2 - 2t$ and $y(t) = t^4 - 8t^2 + 3$, which of the following is $\frac{d^2y}{dx^2}$ in terms of t ?

(A) $\frac{4t+1}{t-2}$

(B) $\frac{4(t+2)}{t-2}$

(C) $\frac{8(t+1)}{t-2}$

(D) $\frac{4t^2-8t}{t-2}$

30. Which of the following definite integral best represents the Riemann sum $\lim_{n\to\infty}\sum_{i=1}^{n}\sin\left(\frac{2i\pi}{n}\right)\frac{\pi}{n}$?

(A) $\int_0^{\frac{\pi}{2}} \sin x\, dx$

(B) $\int_0^{\pi} \sin x\, dx$

(C) $\int_0^{\frac{\pi}{2}} \sin 2x\, dx$

(D) $\int_0^{\pi} \sin 2x\, dx$

END OF PART A OF SECTION I

STOP

CALCULUS BC TEST 1
SECTION I, Part B
Time — 45 minutes
Number of questions — 15

A GRAPHING CALCULATOR IS REQUIRED FOR SOME QUESTIONS ON THIS PART OF THE EXAM.

Directions: Solve each of the following problems using the available space for scratch work. Choose the best answer among the answer choices given and fill in the corresponding circle on the answer sheet.

76. According to the table shown below, all of the following are true about the function f EXCEPT

a	$\lim_{x\to a^-} f(x)$	$\lim_{x\to a^+} f(x)$	$f(a)$
1	2	2	3
2	$-\infty$	∞	undefined
3	1	1	1
4	∞	∞	undefined

(A) f has a hole at $x = 1$

(B) f has vertical asymptotes at $x = 2$ and $x = 4$

(C) f has a horizontal asymptote at $y = 1$

(D) f is continuous at $x = 3$

77. A particle moves in the xy-plane so that the position of the particle at time t is given by $x(t) = e^{-2t}$ and $y(t) = \cos^2 t$. Which of the following is the speed of the particle at $t = 2$?

(A) 1.519 (B) 1.272 (C) 0.922 (D) 0.758

---

78. If the derivative of a function f, f', is given by $f'(x) = \cos(x^3 - 3x^2 - 6x + 15)$, at which of the following values of x does f have a relative minimum on the open interval $(0, 1)$?

(A) 0.135 only

(B) 0.135 and 0.546 only

(C) 0.135 and 0.905 only

(D) 0.135, 0.546, and 0.905

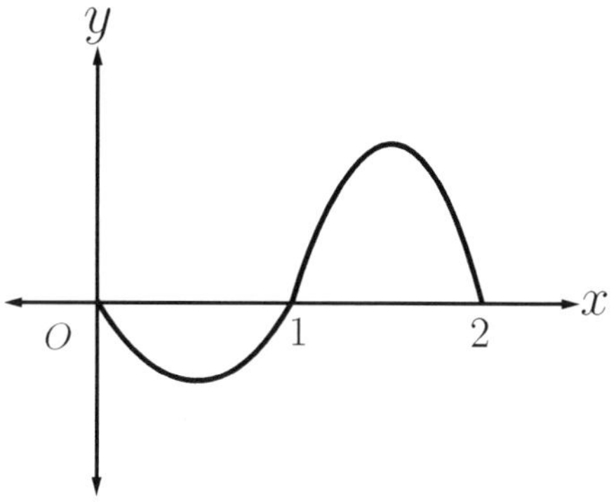

Graph of f'

79. The graph of f', the derivative of f, is shown above. If $f(0) = 1$, which of the following could be the graph of f ?

(A)

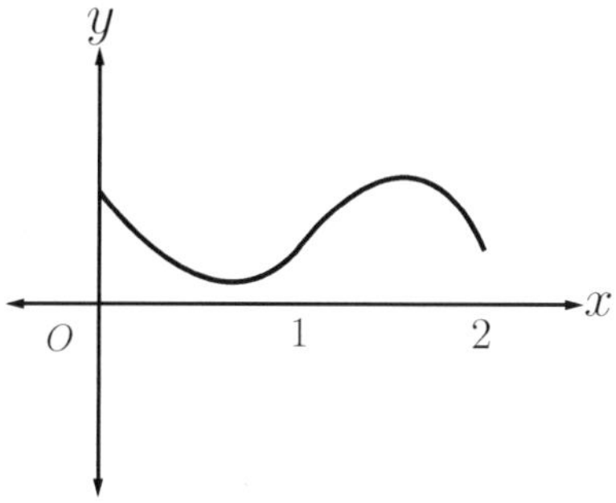

(B)

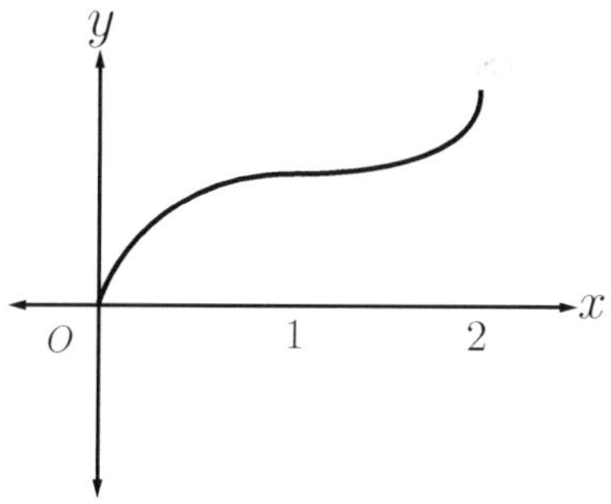

(C)

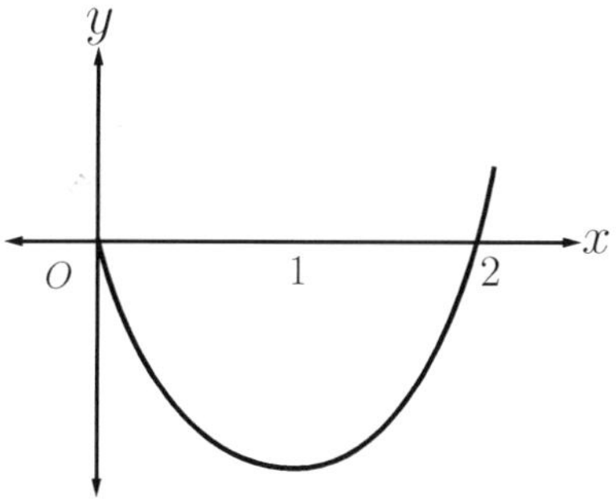

(D)

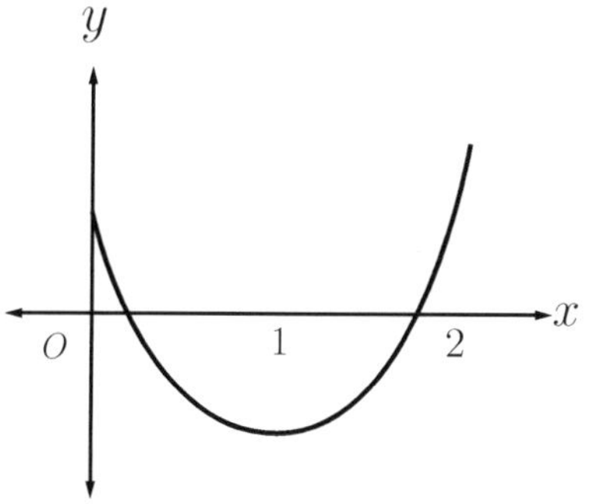

80. If the series $\sum_{n=1}^{\infty} a_n$ and $\sum_{n=1}^{\infty} b_n$, where $a_n > 0$ and $b_n > 0$ for $n \geq 1$ are given and $\sum_{n=1}^{\infty} b_n$ diverges, which of the following must be true?

(A) If $b_n \leq a_n$, then $\sum_{n=1}^{\infty} a_n$ diverges

(B) If $a_n \leq b_n$, then $\sum_{n=1}^{\infty} a_n$ converges

(C) If $\lim_{n\to\infty} \frac{a_n}{b_n} = \frac{1}{2}$, $\sum_{n=1}^{\infty} a_n$ converges

(D) If $\lim_{n\to\infty} \frac{b_n}{a_n} = 0$, $\sum_{n=1}^{\infty} a_n$ diverges

---

81. The region R is enclosed by $y = e^{-x} + 2$, $y = 2x$, and the y-axis. Which of the following is the area of R ?

(A) 1.428 (B) 1.661 (C) 2.252 (D) 2.275

82. If the function f is given by $f(x) = x^4 - 2x^3 - \ln x$, on which of the following interval is the graph of f concave down?

(A) $(-\infty, 0.694)$

(B) $(-0.253, 1.103)$

(C) $(0.586, 0.876)$

(D) $(0.762, 1.361)$

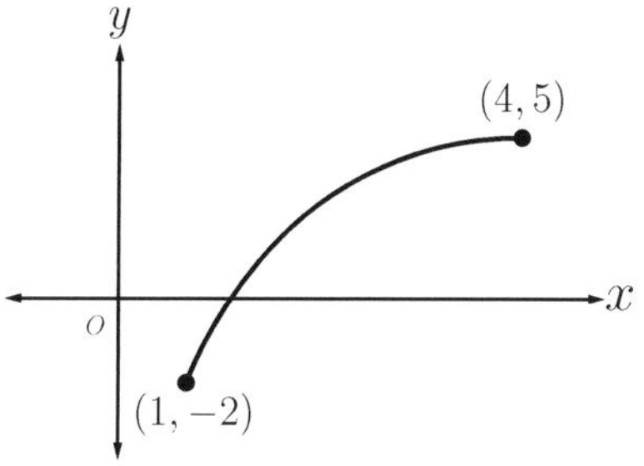

Graph of f

83. A portion of the graph of a differentiable function f is shown above. If c is the value that satisfies the conclusion of the Mean Value Theorem, which of the following must be the slope of the tangent line to the graph of f at $x = c$?

(A) $\frac{7}{3}$ (B) $\frac{5}{3}$ (C) $-\frac{3}{5}$ (D) $-\frac{3}{7}$

84. Coffee is dripping from a filter into a coffee pot at the rate $R(t) = 5.86e^{0.15t}$ mL per minute. At $t = 2$ minutes, the amount of coffee in the pot is 75 mL. Which of the following is the amount of coffee, to the nearest mL, in the pot at $t = 5$ minutes?

(A) 100 mL (B) 105 mL (C) 110 mL (D) 120 mL

---

x	1	2	3	4	5	6	7
$f(x)$	42	38	35	36	40	42	45

85. The table above shows the selected values of a continuous function f on the closed interval $[1, 7]$. Which of the following could be the value of $\int_1^7 f(x)\,dx$ using a midpoint sum with three subintervals of equal length?

(A) 228 (B) 232 (C) 234 (D) 240

86. $\int x^2 \ln x \, dx =$

(A) $\frac{1}{3}x^3 \ln x - \frac{1}{9}x^3 + C$

(B) $\frac{1}{6}x^3 \ln^2 x - \frac{1}{3}x^3 + C$

(C) $\frac{1}{3}x^3 + \frac{1}{2}\ln^2 x + C$

(D) $\frac{1}{3}x^3 + \frac{1}{2}\ln x^2 + C$

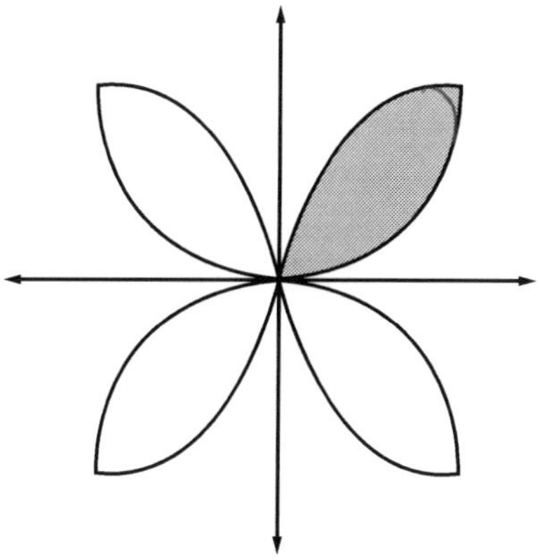

Polar curve $r = \sin 2\theta$

87. The figure above shows the polar curve $r = \sin 2\theta$. Which of the following is the area of the shaded region?

(A) 0.269 (B) 0.393 (C) 0.436 (D) 0.486

88. Which of the following statements are true about the series $\sum_{n=1}^{\infty}(-1)^n a_n$, where $a_n = \dfrac{\sqrt{n}}{n+1}$?

I. $a_{n+1} \leq a_n$

II. $\lim_{n\to\infty} a_n = 0$

III. The series is convergent

(A) I only

(B) II only

(C) I and II only

(D) I, II, and III

---

89. A path of a particle moving in the plane is defined by the parametric equations $x(t) = t\sqrt{t-1}$ and $y(t) = \ln t^2 + 1$. Which of the following is the slope of the tangent line to the path at $t = 5$?

(A) $\dfrac{8}{65}$ (B) $\dfrac{5}{37}$ (C) $\dfrac{3}{13}$ (D) $\dfrac{2}{5}$

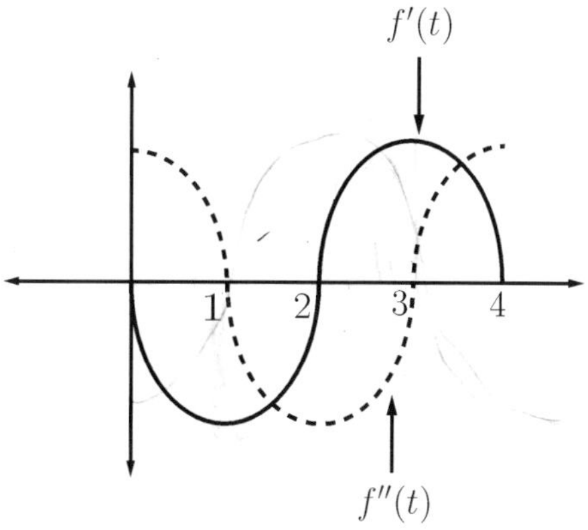

Graphs of $f'(t)$ and $f''(t)$

90. A particle is moving along the x-axis. Let the position of the particle at time t be $f(t)$ for $0 \leq t \leq 4$. If the graphs of $f'(t)$ and $f''(t)$ are shown above, which of the following intervals does the particle speed up?

(A) $1 < t < 3$ only

(B) $2 < t < 4$ only

(C) $1 < t < 2$ and $3 < t < 4$

(D) $0 < t < 2$ and $3 < t < 4$

END OF PART B OF SECTION I

STOP

CALCULUS BC TEST 2

SECTION I, Part A

Time — 60 minutes
Number of questions — 30

A CALCULATOR MAY NOT BE USED ON THIS PART OF THE EXAM.

Directions: Solve each of the following problems using the available space for scratch work. Choose the best answer among the answer choices given and fill in the corresponding circle on the answer sheet.

1. The position vector of a particle moving in the xy-plane is given by $< \ln t, e^{-2t} >$. Which of the following is the acceleration vector of the particle at $t = 3$?

   (A) $< -\frac{1}{3}, \frac{1}{e^6} >$

   (B) $< \frac{1}{3}, -\frac{2}{e^6} >$

   (C) $< -\frac{1}{9}, \frac{4}{e^6} >$

   (D) $< \frac{1}{9}, -\frac{4}{e^6} >$

2. If $y = \left(\dfrac{x+1}{x-1}\right)^3$, then $\dfrac{dy}{dx} =$

(A) $\dfrac{6x(x+1)^2}{(x-1)^4}$

(B) $\dfrac{-6(x+1)^2}{(x-1)^4}$

(C) $\dfrac{-12x(x+1)^2}{(x-1)^5}$

(D) $\dfrac{-12(x+1)^2}{(x-1)^5}$

---

3. For $0 \le \theta \le \pi$, which of the following best represents the total area enclosed by the polar curve $r = 3 - \cos\theta$?

(A) $\dfrac{1}{4}\displaystyle\int_0^{\pi} (3 - \cos\theta)^2 \, d\theta$

(B) $\dfrac{1}{2}\displaystyle\int_0^{\pi} (3 - \cos\theta)^2 \, d\theta$

(C) $\dfrac{1}{2}\displaystyle\int_0^{\pi} (3 - \cos\theta) \, d\theta$

(D) $\displaystyle\int_0^{\pi} (3 - \cos\theta) \, d\theta$

4. $\lim_{x\to 2} \dfrac{\cos\frac{\pi}{x}}{\ln(3-x)} =$

(A) $\dfrac{\pi}{2}$ (B) $\dfrac{\pi}{4}$ (C) $-\dfrac{\pi}{4}$ (D) $-\dfrac{\pi}{2}$

(x, y)	$(1, 2)$	$(1.5, 1.5)$	$(2, 1)$
$\dfrac{dy}{dx}$	-1	-2	3

5. Let $f(x)$ be the solution to the differential equation with initial condition $f(1) = 2$. The table above shows the slope of the tangent line to the graph of $f(x)$ at given points (x, y). Which of the following is the approximation of $f(2)$ using Euler's method, starting at $x = 1$, with two steps of equal size?

(A) $\dfrac{1}{2}$ (B) $\dfrac{3}{2}$ (C) $\dfrac{5}{2}$ (D) $\dfrac{7}{2}$

6. Which of the following series are convergent?

I. $\displaystyle\sum_{n=1}^{\infty} \frac{1}{\sqrt[3]{n}}$ II. $\displaystyle\sum_{n=1}^{\infty} \left(-\frac{5}{4}\right)^{n+1}$ III. $\displaystyle\sum_{n=1}^{\infty} (-1)^n \frac{1}{n}$

(A) III only

(B) I and II only

(C) I and III only

(D) II and III only

---

7. Let g be the function with derivative given by $g'(x) = -2(x+1)(x-2)(x-3)$. Which of the following interval is g decreasing?

(A) $(-\infty, 2]$

(B) $[-1, 3]$

(C) $[-1, 2]$ and $[3, \infty)$

(D) $(-\infty, -1]$ and $[2, 3]$

8. Which of the following best represents the length of the curve given by the parametric equations $x(t) = te^{-t}$ and $y(t) = t^2$ from $t = 0$ to $t = 10$?

(A) $\displaystyle\int_0^{10} \sqrt{e^{-2t}(1-t)^2 + 4t^2}\,dt$

(B) $\displaystyle\int_0^{10} \sqrt{e^{-t}(1-t) + 4t^2}\,dt$

(C) $\displaystyle\int_0^{10} \sqrt{e^{2t}(1-t)^2 + 4t^2}\,dt$

(D) $\displaystyle\int_0^{10} \sqrt{e^{t}(1-t) + 4t^2}\,dt$

---

9. Which of the following is the slope of the tangent line to the graph of $f(x)$ given by $f(x) = \ln(x - x^2)$ at $x = \dfrac{1}{2}$?

(A) $-\dfrac{1}{3}$ (B) 0 (C) $\dfrac{1}{4}$ (D) $\dfrac{1}{3}$

10. Which of the following is the area of a region bounded by $f(x) = \dfrac{3x}{\sqrt{x^2+1}}$, the x-axis, $x = -1$ and $x = 2$?

(A) $\dfrac{3}{2}(\sqrt{7} - 1)$

(B) $\dfrac{1}{2}(\sqrt{5} - \sqrt{3})$

(C) $3(\sqrt{5} - \sqrt{2})$

(D) $3(\sqrt{2} - 1)$

---

11. $\displaystyle\int \frac{x+1}{\sqrt{1-x^2}}\,dx =$

(A) $\sin^{-1} x + \sqrt{1-x^2} + C$

(B) $\sin^{-1} x - \sqrt{1-x^2} + C$

(C) $\tan^{-1} x + 2\sqrt{1-x^2} + C$

(D) $\tan^{-1} x - 2\sqrt{1-x^2} + C$

$f'(x)$	$f''(x)$	$f'''(x)$	$f^{(4)}(x)$
$\frac{1}{2x-1}$	$\frac{-2}{(2x-1)^2}$	$\frac{8}{(2x-1)^3}$	$\frac{-48}{(2x-1)^4}$

12. The table above shows the first, second, third, and fourth order derivatives of f. If $f(1) = 3$, which of the following is the third-degree Taylor polynomial T_3 for f about $x = 1$?

(A) $T_3(x) = 3 + (x-1) - 2(x-1)^2$

(B) $T_3(x) = 3 + (x-1) - (x-1)^2$

(C) $T_3(x) = 3 + (x-1) - 2(x-1)^2 + 8(x-1)^3$

(D) $T_3(x) = 3 + (x-1) - (x-1)^2 + \frac{4}{3}(x-1)^3$

---

13. Consider the series $\sum_{n=1}^{\infty} (-1)^n \frac{n^2(x+2)^n}{3^n}$. Which of the following is the radius of convergence of the series?

(A) 1 (B) $\frac{3}{2}$ (C) 2 (D) 3

x	1	2	4	5	8
$f(x)$	4	9	5	7	10

14. Let f be a continuous function on the closed interval $[1, 8]$. The table shows values of f for selected values of x. Using a trapezoidal rule with the intervals $[1, 2]$, $[2, 4]$, $[4, 5]$, and $[5, 8]$, which of the following is the approximation of $\int_1^8 f(x)\,dx$?

(A) 104 (B) 52 (C) 36 (D) 28

---

15. If the function f is continuous and given by $f(x) = xe^{x^2}$, which of the following is the average value of f on the closed interval $[0, 5]$?

(A) $\frac{1}{5}(e^5 - 1)$ (B) $\frac{1}{10}(e^5 - e)$ (C) $\frac{1}{10}(e^{25} - 1)$ (D) $\frac{1}{5}(e^{25} - e)$

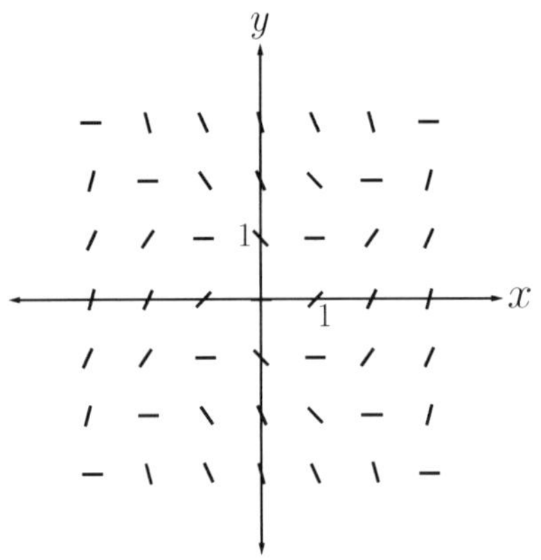

16. Which of the following differential equation gives a slope field shown above?

(A) $\dfrac{dy}{dx} = x^2 + y^2$

(B) $\dfrac{dy}{dx} = x^2 - y^2$

(C) $\dfrac{dy}{dx} = x + y$

(D) $\dfrac{dy}{dx} = x - y$

---

17. Which of the following is an equation of the tangent line to the curve defined by $x(t) = t^3 - 2t$ and $y(t) = 2t^2 - 3t + 1$ at $t = 2$?

(A) $y = 2x + \dfrac{1}{2}$

(B) $y = 2x - \dfrac{3}{2}$

(C) $y = \dfrac{1}{2}x + 1$

(D) $y = \dfrac{1}{2}x - 3$

18. $\displaystyle\int_1^e \ln x\,dx =$

(A) 1 (B) $\frac{1}{2}$ (C) $\frac{1}{e}$ (D) $\frac{1-e}{e}$

---

19. If $x^2y + xy^2 = 6x$, then which of the following is the value of $\frac{dy}{dx}$ when $x = 1$ and $y = 2$?

(A) $\frac{3}{4}$ (B) $\frac{2}{7}$ (C) $-\frac{1}{3}$ (D) $-\frac{2}{5}$

20. Let T_2 be the second-degree Taylor polynomial for $f(x) = \ln\left(\frac{x}{2}\right)$ about $x = 2$. What is approximation of $\ln 2$ using T_2 ?

(A) $\frac{1}{3}$ (B) $\frac{1}{2}$ (C) $\frac{3}{5}$ (D) $\frac{4}{7}$

---

21. Consider the power series $\sum_{n=1}^{\infty} \frac{2^n(2x-4)^n}{n+1}$. The series converges at the following values of x EXCEPT

(A) $\frac{7}{4}$ (B) $\frac{15}{8}$ (C) 2 (D) $\frac{5}{2}$

22. $\lim_{h\to 0} \dfrac{\ln(e+h) - \ln e}{h} =$

(A) $\dfrac{1}{2e}$ (B) $\dfrac{1}{e}$ (C) e (D) $2e$

---

23. Which of the following series converges?

I. $\displaystyle\sum_{n=1}^{\infty} \frac{1}{n^{\pi}}$ II. $\displaystyle\sum_{n=1}^{\infty} ne^{-n2}$ III. $\displaystyle\sum_{n=1}^{\infty} \frac{10n - n^3}{3n^3 + 2n^2}$

(A) II only

(B) III only

(C) I and II only

(D) I and III only

24. If the function f is given by $f(x) = \dfrac{1}{(1-x)^2}$, which of the following is the coefficient of x^2 in the Taylor series for f about $x = 0$?

(A) 4 (B) 3 (C) $\dfrac{1}{8}$ (D) 0

---

25. Let R be the unbounded region between the graphs of $y = \dfrac{1}{x^2+1}$, the line $x = 1$, and x-axis for $x \geq 1$. What is the area of R ?

(A) π (B) $\dfrac{3\pi}{4}$ (C) $\dfrac{\pi}{2}$ (D) $\dfrac{\pi}{4}$

26. $\displaystyle\int \frac{x-12}{x^2-3x-10}\,dx =$

(A) $2\ln|x+2| - \ln|x-5| + C$

(B) $2\ln|x+2| + \ln|x-5| + C$

(C) $\ln|x+2| - 2\ln|x-5| + C$

(D) $\ln|x+2| + 2\ln|x-5| + C$

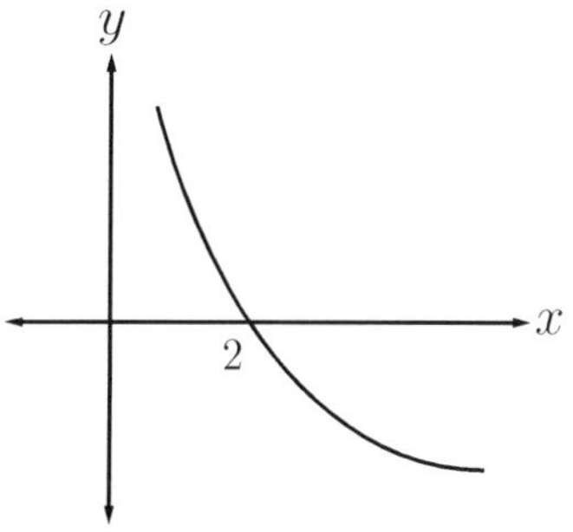

27. A portion of the graph of the function f is shown above. Which of the following statement must be true?

(A) $f(2) < f''(2) < f'(2)$

(B) $f''(2) < f'(2) < f(2)$

(C) $f'(2) < f(2) < f''(2)$

(D) $f(2) < f'(2) < f''(2)$

28. $\lim_{x\to\infty}(xe^{\frac{1}{x}} - x) =$

(A) e (B) 2 (C) 1 (D) $\frac{1}{e}$

---

29. If the function f is given by $f(x) = x^3 - 6x^2 + 11x - 7$, which of the following values of x does the instantaneous rate of change of f equal the average rate of change of f on the closed interval $[0, 4]$?

(A) $x = \frac{6 \pm 2\sqrt{2}}{3}$

(B) $x = \frac{6 \pm 2\sqrt{3}}{3}$

(C) $x = \frac{2 \pm \sqrt{2}}{3}$

(D) $x = \frac{2 \pm \sqrt{3}}{3}$

30. A particle moves along a straight line with its velocity at time t is given by $v(t) = t^2 - 4t + 3$. Which of the following is the total distance the particle moves from $t = 0$ and $t = 3$?

(A) $\dfrac{11}{3}$ (B) $\dfrac{10}{3}$ (C) 3 (D) $\dfrac{8}{3}$

END OF PART A OF SECTION I

STOP

CALCULUS BC TEST 2
SECTION I, Part B
Time — 45 minutes
Number of questions — 15

A GRAPHING CALCULATOR IS REQUIRED FOR SOME QUESTIONS ON THIS PART OF THE EXAM.

Directions: Solve each of the following problems using the available space for scratch work. Choose the best answer among the answer choices given and fill in the corresponding circle on the answer sheet.

76. Let f and g be the functions. The table below shows values of f, g, f', and g'. If the function h is defined by $h(x) = \dfrac{g(x)}{f(x)}$, which of the following is the value of $h'(3)$?

x	f	g	f'	g'
3	2	-1	3	-4

(A) $\frac{5}{4}$ (B) $\frac{3}{4}$ (C) $-\frac{3}{4}$ (D) $-\frac{5}{4}$

77. Water is leaking out of a tank at the rate $L(t) = 275(1.1)^t$ at the same time water is being pumped into the tank at the rate $R(t) = 300\cos(t-3)+350$, where t is measured in minutes. The amount of water in the tank at $t = 0$ minute is 250 cm^3. For $0 \le t \le 8$, at what time does the amount of water in the tank reach its maximum?

(A) 6.711 minutes

(B) 5.593 minutes

(C) 4.348 minutes

(D) 2.984 minutes

---

x	0	1	4	5	7
$f(x)$	3	-2	1	5	3

78. Let f be a continuous function on the closed interval $[0, 9]$. The values of f for selected values of x are shown in the table above. Which of the following must be true?

(A) $\int_0^1 f'(x)\,dx = -5$

(B) $\int_0^4 f'(x)\,dx = 1$

(C) $\int_1^5 f'(x)\,dx = 3$

(D) $\int_4^7 f'(x)\,dx = -2$

79. Let f be a continuous function on the closed interval $[2,5]$. If $f(2) = -1$ and $f(5) = 8$, which of the following must be true about f ?

(A) There is at least one number c in the interval $(2,5)$ such that $f'(c) = 3$.

(B) There at least one zero of f in the interval $(2,5)$.

(C) f is increasing in the interval $[2,5]$.

(D) f is concave up in the interval $[2,5]$.

---

80. A particle moves along the x-axis. The position of the particle at time t is given by $x(t) = t^4 - 2t^3 - 9t^2 + 2t + 8$. For $0 \le t \le 3$, which of the following is the velocity of the particle when the position of the particle is 1 unit to the right of the origin?

(A) -18.263 (B) -16.955 (C) 8.327 (D) 12.692

81. Which of the following series is absolutely convergent?

(A) $\sum_{n=1}^{\infty}(-1)^n \frac{1}{n}$

(B) $\sum_{n=1}^{\infty} \frac{\cos n\pi}{n^2}$

(C) $\sum_{n=1}^{\infty} \frac{3^n}{2^{n+1}}$

(D) $\sum_{n=1}^{\infty}(-1)^{n+1} \frac{n^2-3n}{3n^2+1}$

---

82. Let f be the function with the first derivative f' given by $f'(x) = x^3 - 8x^2 + 19x - 11$. Which of the following statements are true about f in the interval $[0, 4]$?

I. f has a relative minimum in the interval $[0, 4]$.
II. f has a relative maximum in the interval $[0, 4]$.
III. f is increasing in the interval $[0, 4]$.

(A) I only

(B) II only

(C) I and III only

(D) II and III only

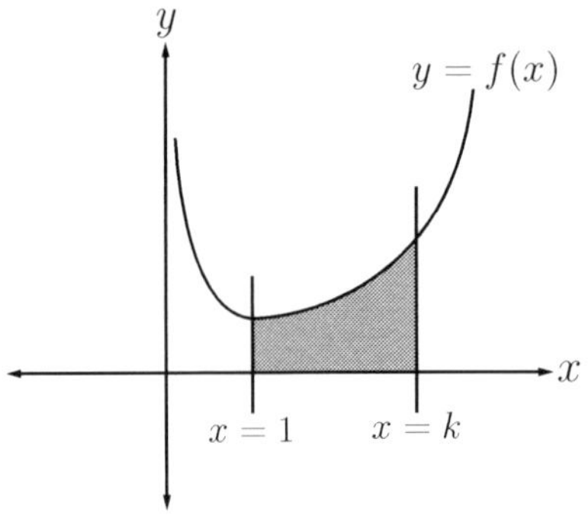

83. The shaded region R shown above is bounded by the graph of the function $f(x) = \dfrac{e^x}{x}$, the lines, $x = 1$ and $x = k$, and the x-axis. If k increases at a rate of 4 units per minute, how fast is the area of R increasing when $k = 2$?

(A) $8e$ (B) $\dfrac{e}{2}$ (C) $\dfrac{e^2}{4}$ (D) $2e^2$

---

84. Which of the following is the slope of the tangent line to the polar curve $r = 2\sin\theta$ at $\theta = \dfrac{2\pi}{3}$?

(A) $\sqrt{3}$ (B) $\sqrt{2}$ (C) $\dfrac{\sqrt{2}}{2}$ (D) $\dfrac{\sqrt{3}}{3}$

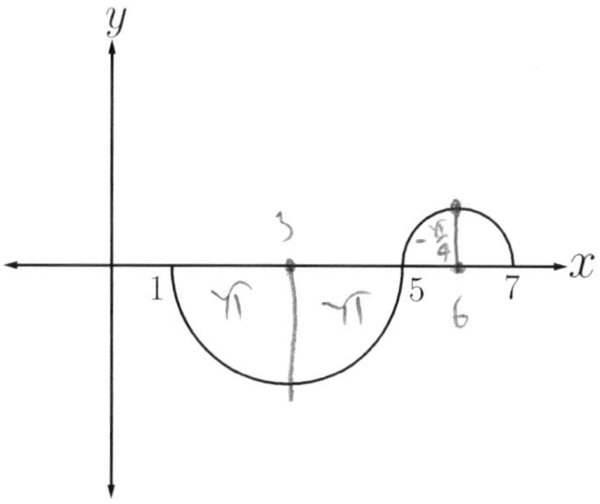

Graph of f

85. The graph of the function f shown above consists of two semicircles with radii 2 and 1. If $g(x) = 3x + \int_6^x f(t)\,dt$, then which of the following is the value of $g(1)$?

(A) $-\frac{3\pi}{2} - 3$ (B) $\frac{7\pi}{4} - 3$ (C) $3 + \frac{3\pi}{2}$ (D) $3 + \frac{7\pi}{4}$

---

86. The velocity of a particle moving along the y-axis is given by $v(t) = e^t \sin t$ for $0 \le t \le 4$. What is the total distance traveled by the particle from $t = 0$ to $t = 4$?

(A) -2.316 (B) 15.374 (C) 26.457 (D) 31.272

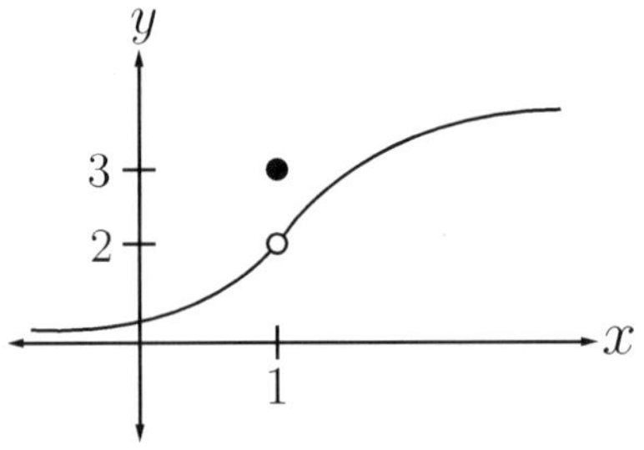

Graph of f

87. If the graph of the function f is shown above, which of the following is the value of $\lim_{x \to 1} f(2-x)$?

(A) 1 (B) 2 (C) 3 (D) nonexistent

---

88. A water tank has the shape of an inverted circular cone with base radius 8 cm and height 20 cm. Water is being pumped into the tank at a rate of 12 cm^3/min. What is the rate at which the radius of the water surface is increasing when the water is 5 cm deep?

(A) $\dfrac{3}{5\pi}$ cm/min

(B) $\dfrac{6}{5\pi}$ cm/min

(C) $\dfrac{9}{25\pi}$ cm/min

(D) $\dfrac{12}{25\pi}$ cm/min

89. The position of a particle moving in the x-axis at time t is given by $x(t) = \frac{1}{3}t^3 - 2t^2 - 5t + 6$. Which of the following statements are true about the particle at $t = 3$?

I. The particle is 18 units to the left of the origin.

II. The particle is moving to the right.

III. The particle is moving at a decreasing speed.

(A) I only

(B) II only

(C) I and III only

(D) II and III only

90. The meerkat population of an isolated island at time t is modeled by the logistic differential equation $\frac{dP}{dt} = \frac{1}{10}P - \frac{P^2}{50000}$. If the initial meerkat population is 1000, which of the following is the solution to the logistic differential equation?

(A) $P(t) = \dfrac{1000}{1 + 0.8e^{-0.1t}}$

(B) $P(t) = \dfrac{5000}{1 + 0.8e^{-0.1t}}$

(C) $P(t) = \dfrac{5000}{1 + 2e^{-0.1t}}$

(D) $P(t) = \dfrac{5000}{1 + 4e^{-0.1t}}$

END OF PART B OF SECTION I

STOP

CALCULUS BC TEST 3
SECTION I, Part A
Time — 60 minutes
Number of questions — 30

A CALCULATOR MAY NOT BE USED ON THIS PART OF THE EXAM.

Directions: Solve each of the following problems using the available space for scratch work. Choose the best answer among the answer choices given and fill in the corresponding circle on the answer sheet.

1. Which of the following gives the length of the polar curve defined by $r = 1 + 2\sin\theta$ for $0 \le \theta \le \dfrac{\pi}{2}$?

(A) $\displaystyle\int_0^{\frac{\pi}{2}} \sqrt{(1+2\sin\theta)^2 + \cos^2\theta}\, d\theta$

(B) $\displaystyle\int_0^{\frac{\pi}{2}} \sqrt{(1+2\sin\theta)^2 + 4\cos^2\theta}\, d\theta$

(C) $\displaystyle\int_0^{\frac{\pi}{2}} \sqrt{(1+2\sin\theta) + 2\cos\theta}\, d\theta$

(D) $\displaystyle\int_0^{\frac{\pi}{2}} \sqrt{(1+2\sin\theta) + 4\cos\theta}\, d\theta$

2. Let $y = f(x)$ be the solution to the differential equation $\dfrac{dy}{dx} = \dfrac{1}{2}(x + y)$ with initial condition $f(2) = 4$. What is the approximation of $f(3)$ using Euler's method, starting at $x = 2$, with two steps of equal length?

(A) 6 (B) 6.5 (C) 7 (D) 7.5

---

3. Which of the following series converges?

(A) $\displaystyle\sum_{n=1}^{\infty} \frac{2n}{\sqrt{n+1}}$

(B) $\displaystyle\sum_{n=1}^{\infty} \frac{n}{n^2+1}$

(C) $\displaystyle\sum_{n=1}^{\infty} \frac{3n}{n^3+2n}$

(D) $\displaystyle\sum_{n=1}^{\infty} \frac{n^3+1}{4n^3+3n}$

4. If the functions f and g are defined by $f(x) = \sqrt{x}$ and $g(x) = x^2 + x$, which of the following is the derivative of $f(g(x))$?

(A) $\dfrac{x+1}{\sqrt{x^2+x}}$

(B) $\dfrac{2x+1}{2\sqrt{x^2+x}}$

(C) $\dfrac{x+1}{x^2+x}$

(D) $\dfrac{2x+1}{x^2+x}$

---

5. If $\sin(xy) = 2xy$, then $\dfrac{dy}{dx} =$

(A) $\dfrac{2y - y\cos(xy)}{x\cos(xy) - 2x}$

(B) $\dfrac{2y - x\cos(xy)}{y\cos(xy) - 2x}$

(C) $\dfrac{x - y\cos(xy)}{x\cos(xy) - y}$

(D) $\dfrac{y - x\cos(xy)}{y\cos(xy) - x}$

6. Let g be the function defined by $g(x) = \int_1^{x^2} 2t - 1\, dt$. What is the slope of the tangent line to the graph of g at $x = 2$?

(A) 6 (B) 15 (C) 21 (D) 28

---

7. $\displaystyle\int \frac{4}{x^2 + 2x - 3}\, dx =$

(A) $\ln|x^2 + 2x - 3| + C$

(B) $\ln\left|\dfrac{2x+2}{x^2 + 2x - 3}\right| + C$

(C) $\ln\left|\dfrac{x-3}{x+1}\right| + C$

(D) $\ln\left|\dfrac{x-1}{x+3}\right| + C$

x	0	5	15	25	30
$f(x)$	3	9	5	2	6

8. Let f be a continuous function on the closed interval $[0, 30]$. The values of f for selected values of x are given in the table above. What is approximation of $\int_0^{30} f(x)\,dx$ using right Riemann sum with 4 subintervals $[0, 5]$, $[5, 15]$, $[15, 25]$, and $[25, 30]$?

(A) 115 (B) 135 (C) 145 (D) 165

---

9. $\int_0^{\frac{\pi}{2}} \sin^2 x\,dx =$

(A) $\frac{\pi}{4}$ (B) $\frac{\pi}{2}$ (C) $\frac{3\pi}{4}$ (D) π

10. A particle moves in the xy-plane so that its position vector is $< \frac{1}{t+1}, e^{1-t} >$. As the particle passes through the point $\left(\frac{1}{2}, 1\right)$, which direction does the particle move?

(A) The particle moves to the right and up.

(B) The particle moves to the right and down.

(C) The particle moves to the left and up.

(D) The particle moves to the left and down.

---

11. Which of the following is the derivative of $y = x^x$?

(A) $\frac{dy}{dx} = x^x(x \ln x + 1)$

(B) $\frac{dy}{dx} = x^x(\ln x + 1)$

(C) $\frac{dy}{dx} = x^x \cdot \ln x$

(D) $\frac{dy}{dx} = x \cdot x^{x-1}$

12. Let f be a function with first derivative $f'(x) = \dfrac{1}{\sqrt{1-2x}}$. Which of the following is the coefficient of x^3 in the Taylor series for f about $x = 0$?

(A) -3 (B) -2 (C) $\dfrac{1}{3}$ (D) $\dfrac{1}{2}$

---

13. Which of the following is the radius of convergence for the power series $\displaystyle\sum_{n=1}^{\infty}(-1)^n \frac{(x-3)^n}{n \cdot 5^n}$?

(A) 6 (B) 5 (C) 3 (D) $\dfrac{1}{5}$

14. If $x(t) = \sin t - 1$ and $y(t) = \cos t + 1$, then what is $\dfrac{d^2y}{dx^2}$ in terms of t ?

(A) $-\tan^2 t$ (B) $-\sin^2 2t$ (C) $-\csc^3 t$ (D) $-\sec^3 t$

---

15. $\lim_{x \to 0^+} x \ln x =$

(A) e (B) 1 (C) $\dfrac{1}{2}$ (D) 0

16. $\displaystyle\int \tan^2 x - 3^x \, dx =$

(A) $\csc x - x - 3^x \cdot \ln 3 + C$

(B) $\sec x + x - 3^x \cdot \ln 3 + C$

(C) $\tan x - x - \dfrac{3^x}{\ln 3} + C$

(D) $\cot x + x - \dfrac{3^x}{\ln 3} + C$

---

17. Which of the following is the sum of the series $1 - 2 + \dfrac{2^2}{2!} - \dfrac{2^3}{3!} + \cdots + (-1)^n \dfrac{2^n}{n!} + \cdots$?

(A) $\cos 2$ (B) $\ln 2$ (C) $\dfrac{1}{2}$ (D) $\dfrac{1}{e^2}$

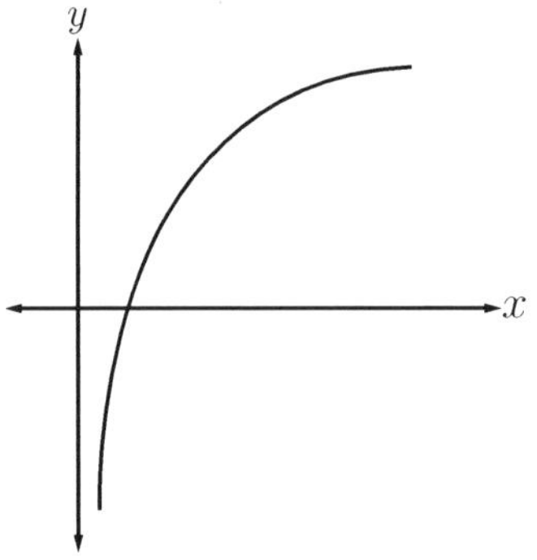

Graph of f

18. The graph of the function f is shown above. Which of the following could be the graph of f' ?

(A)

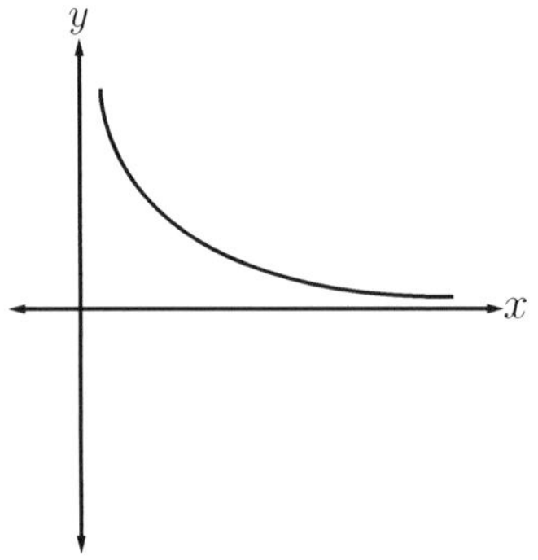

(B)

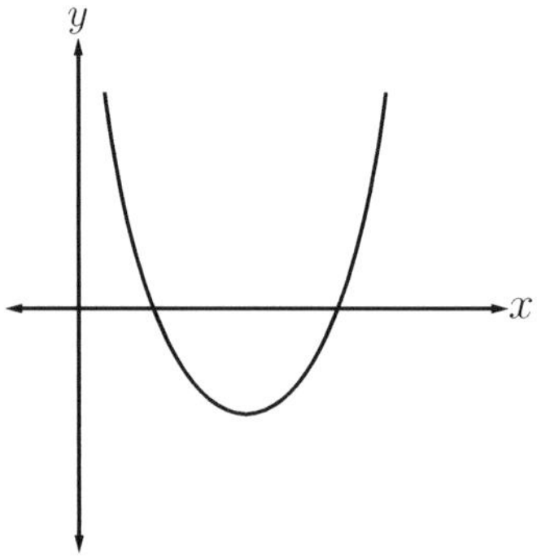

(C)

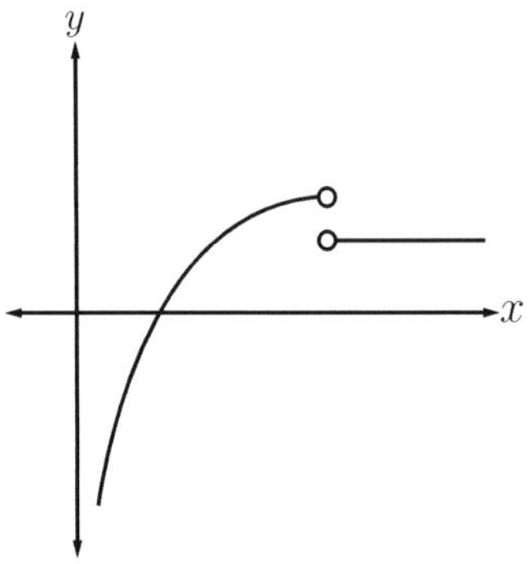

(D)

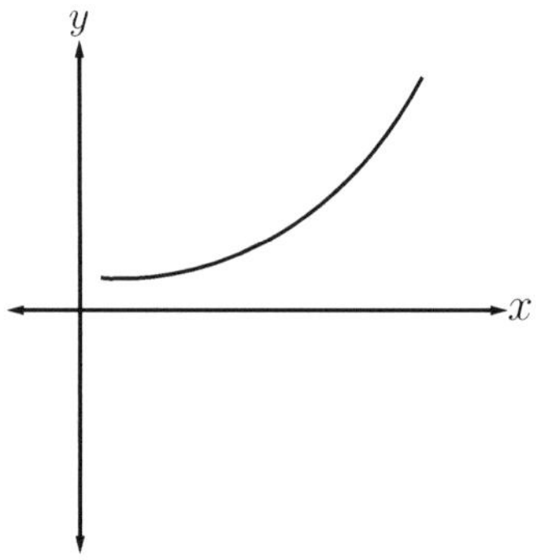

19. Which of the following is the largest area of a rectangle whose base on the x-axis and its other two vertices above the x-axis and lying on the parabola $y = 9 - x^2$?

(A) 28 (B) 24 (C) $12\sqrt{3}$ (D) $9\sqrt{2}$

---

20. If $x(0) = \frac{1}{e}$ and $\frac{dx}{dt} = te^{-t}$, which of the following is the value of $x(1)$?

(A) $\frac{e+3}{e}$ (B) $\frac{e+2}{e}$ (C) $\frac{e+1}{e}$ (D) $\frac{e-1}{e}$

21. $\lim_{h\to 0} \dfrac{(-1+h)^4 - (-1)^4}{h} =$

(A) -6 (B) -4 (C) 2 (D) 3

22. If the function f is continuous on the closed interval $[a, b]$ and is differentiable on the open interval (a, b), which of the following statements must be true?

I. If $f(a) = f(b)$, then there must be c such that $f'(c) = 0$, where $a < c < b$.

II. There must be c such that $(b - a)f'(c) = f(b) - f(a)$, where $a < c < b$.

III. There must be c such that $f(c) = 0$, where $a < c < b$.

(A) I only

(B) III only

(C) I and II only

(D) II and III only

23. If $\int_1^{\infty} \frac{1}{x^p}\,dx = 2$, which of the following must be true?

(A) $\sum_{n=1}^{\infty} \frac{1}{n^p}$ diverges

(B) $\sum_{n=1}^{\infty} \frac{1}{n^p}$ converges

(C) $\sum_{n=1}^{\infty} \frac{1}{n^{p+1}}$ diverges

(D) $\sum_{n=1}^{\infty} \frac{1}{n^{p+1}}$ converges

---

24. The function f is defined by $f(x) = \ln x + e^x$. If the function g is the inverse function of f, which of the following is the value of $g'(e)$?

(A) $\frac{1}{1+e}$ (B) $\frac{1}{e}$ (C) $e-1$ (D) $2e$

25. If the curve is defined by the parametric equations $x(t) = t^3 - 4t^2 - 3t$ and $y(t) = t^4 - 8t^2 + 3$, at what values of t does the curve have a vertical tangent?

(A) $t = 3$ only

(B) $t = -\dfrac{1}{3}$ and $t = 3$

(C) $t = -2$ only

(D) $t = -2$ and $t = 2$

---

26. Let $P = f(t)$ be the solution to the differential equation $\dfrac{dP}{dt} = kP$, where $k \neq 0$. Which of the following could be P ?

(A) $P(t) = \dfrac{1}{3}e^{-kt}$

(B) $P(t) = 3e^{-kt} + 2$

(C) $P(t) = \dfrac{1}{3}e^{kt}$

(D) $P(t) = 3e^{kt} + 2$

27. If the function g is defined by $g(x) = \int_{-2}^{x} t^3 - 6t^2 + 9t\,dt$, which of the following interval is g concave up?

(A) $2 \leq x \leq 4$

(B) $0 \leq x \leq 2$ and $x \geq 4$

(C) $x \leq 1$ and $x \geq 3$

(D) $1 \leq x \leq 3$

---

28. If the region R is bounded by the graphs of $x = \frac{1}{2}y^2 - 2$ and $x = -\frac{1}{2}y^2 + 2$, which of the following is the area of R ?

(A) $\frac{35}{3}$ (B) $\frac{32}{3}$ (C) $\frac{28}{3}$ (D) $\frac{25}{3}$

29. The derivative function of f, f', is defined by $f'(x) = g(x)(x^2 - 4x + 3)$, where g is a differentiable function such that $g(x) < 0$ for all real numbers x. Which of the following must be true about f ?

(A) f has a relative maximum at $x = 1$ and a relative minimum at $x = 3$.

(B) f has a relative minimum at $x = 1$ and a relative maximum at $x = 3$.

(C) f has a relative maximum at $x = 2$.

(D) f has a relative minimum at $x = 2$.

---

30. Let f be a function that has derivatives of all orders for all real numbers x. The Taylor series for f about $x = 0$ converges at $x = 1$, and $\left|f^{(n)}(x)\right| \le \dfrac{n+1}{n^2}$ for $1 \le n \le 4$. If $P_3(x)$ is the third-degree Taylor polynomial of f about $x = 0$, which of the following is the maximum error, $|f(1) - P_3(1)|$, guaranteed by the Lagrange error bound?

(A) $\dfrac{4}{9}$

(B) $\dfrac{5}{16}$

(C) $\dfrac{4}{9 \cdot 3!}$

(D) $\dfrac{5}{16 \cdot 4!}$

END OF PART A OF SECTION I

STOP

CALCULUS BC TEST 3
SECTION I, Part B
Time — 45 minutes
Number of questions — 15

A GRAPHING CALCULATOR IS REQUIRED FOR SOME QUESTIONS ON THIS PART OF THE EXAM.

Directions: Solve each of the following problems using the available space for scratch work. Choose the best answer among the answer choices given and fill in the corresponding circle on the answer sheet.

76. The region R is bounded by the graphs of the functions $f(x) = 50 + 10\sin\left(\frac{\pi}{12}x - 3\right)$, the y-axis, and $g(x) = 20e^{0.4x}$ for $x \geq 0$. Which of the following is the area of R ?

(A) 30.973 (B) 32.246 (C) 33.586 (D) 35.547

x	$f(x)$	$f'(x)$	$f''(x)$	$f'''(x)$
1	2	-4	6	-8

77. The function f has derivatives of all orders for all real numbers. The values of nth order derivatives at $x = 1$ are shown in the table above. If the function g is defined by $g(x) = \int_1^x f(t)\,dt$, which of the following is the third-degree Taylor polynomial for g about $x = 1$?

(A) $-2(x-1) + (x-1)^2 + 2(x-1)^3$

(B) $-2(x-1) - (x-1)^2 - 2(x-1)^3$

(C) $2(x-1) + 2(x-1)^2 - (x-1)^3$

(D) $2(x-1) - 2(x-1)^2 + (x-1)^3$

---

78. If the function f is defined by $f(x) = x\ln(x-1)$, what is the average value of f on the closed interval $[2, 6]$?

(A) 3.282 (B) 4.541 (C) 5.464 (D) 6.681

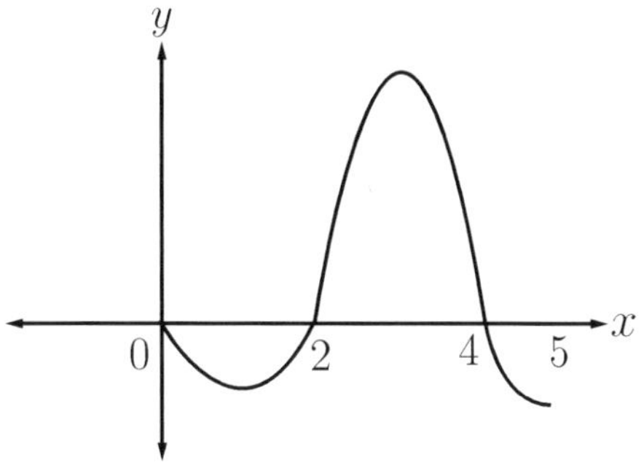

Graph of f'

79. The graph of f', the derivative of the function f, for $0 \leq x \leq 5$ is shown above. At what value of x does f attain the absolute maximum?

(A) 0 (B) 2 (C) 4 (D) 5

---

80. Let f be a twice-differential function. If the graph of f lies above all of its tangent lines on the open interval $(1, 7)$, which of the following statements must be true?

(A) $f(2) > 0$

(B) $f'(3) < 0$

(C) $f''(5) > 0$

(D) $f''(6) < 0$

x	1	2	4
f	2	1	-3
g	-1	-3	4
f'	3	2	-1
g'	-2	3	1

81. The table above shows values of f, g, f', and g' for selected values of x. If the function h is defined by $h(x) = f(x) \cdot g(2x)$, which of the following is the value of $h'(2)$?

(A) -3 (B) 4 (C) 9 (D) 10

---

82. Let f be the function with $f(-2) = 3$ and $f(2) = -3$. The graph of f is symmetric about the origin. If f is continuous on the interval $[-2, 2]$ and is differentiable on the interval $(-2, 2)$, all of the following are true EXCEPT

(A) $f'(x) < 0$ and $f''(x) > 0$

(B) There must be c in $(-2, 2)$ such that $f(c) = 0$

(C) There must be c in $(-2, 2)$ such that $f'(c) = -\dfrac{3}{2}$

(D) $\displaystyle\int_{-2}^{2} f(x)\,dx = 0$

83. The base of a region is enclosed by $y \geq \frac{1}{2}x^2$ and $y \leq 2$. Which of the following is the volume of the solid if cross-sections perpendicular to the x-axis are semicircles?

(A) 3.351 (B) 4.272 (C) 5.696 (D) 6.421

---

84. The rate at which patients entering a medical clinic is modeled by $R(t) = 20\cos(x - 2) + 30$ for $t \geq 0$. $R(t)$ is measured in the number of patients per hour. If the number of patients in the clinic at $t = 9$ is 265, approximately which of the following is the number of patients in the clinic at $t = 4$?

(A) 385 (B) 224 (C) 120 (D) 106

85. Let $a_n = \dfrac{(-1)^n + 2^n}{4^n}$. Which of the following is the sum of the infinite series $\sum_{n=1}^{\infty} a_n$?

(A) $\frac{5}{6}$ (B) $\frac{4}{5}$ (C) $\frac{5}{7}$ (D) Divergent

---

86. Which of the following series are conditionally convergent?

I. $\sum_{n=1}^{\infty} \dfrac{\cos n\pi}{n}$ II. $\sum_{n=1}^{\infty} \dfrac{(-1)^n}{n^2}$ III. $\sum_{n=1}^{\infty} \dfrac{(-1)^n 2^n}{n!}$

(A) I only

(B) II only

(C) I and III only

(D) II and III only

87. If f', the derivative of the function f, is defined by $f'(x) = x^2 \cos(3x)$, at which of the following value of x does f attain a relative maximum for $1 \leq x \leq 3$?

(A) 1.571 (B) 2.187 (C) 2.618 (D) 2.889

---

88. If the function f is defined by $f(x) = 2\ln(x^2 + 4)$, what values of x satisfy the conclusion of the Mean Value Theorem on the closed the interval $[1, 5]$?

(A) $x = 2.274$ only

(B) $x = 3.662$ only

(C) $x = 1.681$ and $x = 4.282$

(D) $x = 1.190$ and $x = 3.360$

89. A man walks along a straight path at a speed of 10 ft/s. A searchlight is located on the ground 50 ft from the path and is kept focused on the man. At what rate is the searchlight rotating when the man is 120 ft from the point on the path closest to the searchlight?

(A) 0.0172 rad/sec

(B) 0.0296 rad/sec

(C) 0.0363 rad/sec

(D) 0.0429 rad/sec

---

90. Which of the following are the values of x for which the power series $\sum_{n=1}^{\infty} \frac{2^n(x-3)^n}{\sqrt{n+1}}$ converges?

(A) $\frac{5}{2} \leq x < \frac{7}{2}$

(B) $\frac{5}{2} \leq x \leq \frac{7}{2}$

(C) $5 < x < 7$

(D) $5 < x \leq 7$

END OF PART B OF SECTION I

STOP

CALCULUS BC TEST 4
SECTION I, Part A
Time — 60 minutes
Number of questions — 30

A CALCULATOR MAY NOT BE USED ON THIS PART OF THE EXAM.

Directions: Solve each of the following problems using the available space for scratch work. Choose the best answer among the answer choices given and fill in the corresponding circle on the answer sheet.

$$f(x) = \begin{cases} \sqrt{x+k}, & x \geq 2 \\ x^2 - 1, & x < 2 \end{cases}$$

1. If the function f is defined above, which of the following is the value of k so that f is continuous at $x = 2$?

(A) 3 (B) 4 (C) 5 (D) 7

2. $\displaystyle\int \frac{1}{x^2+2x}\,dx =$

(A) $\dfrac{1}{2}\ln\left|\dfrac{x}{x+2}\right| + C$

(B) $\dfrac{1}{2}\ln\left|\dfrac{x+2}{x}\right| + C$

(C) $2\ln\left|\dfrac{1}{x^2+2x}\right| + C$

(D) $2\ln|x^2+2x| + C$

---

3. Which of the following is the sum of the infinite series $\displaystyle\sum_{n=0}^{\infty}(-1)^n\frac{3^n}{e^n}$?

(A) $\dfrac{e}{e+3}$ (B) $\dfrac{e-3}{e}$ (C) $\dfrac{e-1}{e}$ (D) ∞

4. $\displaystyle\int \frac{x^3+2x+1}{x}\,dx =$

(A) $\frac{1}{3}x^3 + 2\ln|x+1| + C$

(B) $\frac{1}{3}x^3 + 2x + \ln|x| + C$

(C) $\frac{1}{4}x^4 + x^2 + x + C$

(D) $\frac{1}{4}x^4 + x^2 + \ln|x-1| + C$

---

5. If the first derivative of f, f', is defined by $f'(x) = -(x+2)(x-1)(x-3)^2$, at what values of x does f has a local minimum?

(A) -2, 1, and 3

(B) -2 and 3 only

(C) 1 and 3 only

(D) -2 only

6. $\displaystyle\lim_{x\to-\infty} \frac{2x}{\sqrt{x^2+1}} =$

(A) -2 (B) 0 (C) 2 (D) nonexistent

---

7. Which of the following is the slope of the tangent line to the graph of $y = \sin x \cos x$ at $x = \dfrac{\pi}{12}$?

(A) $\dfrac{\sqrt{3}}{2}$ (B) $\dfrac{\sqrt{2}}{2}$ (C) $\dfrac{1}{2}$ (D) 0

8. Let $S_n = \left(\frac{1}{n} - \frac{1}{n+1}\right)$. Which of the following is the sum of the series $\lim_{k\to\infty} \sum_{n=1}^{k} S_n$?

(A) 1 (B) $\frac{1}{2}$ (C) $\frac{1}{3}$ (D) divergent

---

9. $\int_{-1}^{1} \frac{1}{\sqrt{2-2x}}\,dx =$

(A) $\frac{1}{2}$ (B) 1 (C) $\frac{3}{2}$ (D) 2

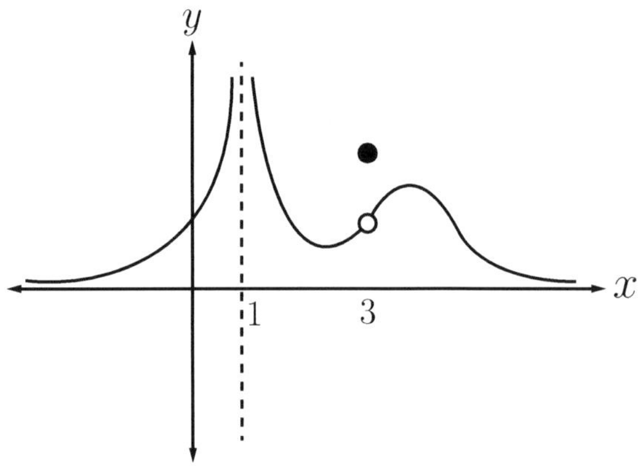

Graph of f

10. The graph of the function f is shown above. Which of the following statements are true?

I. $\lim_{x \to 1} f(x)$ exists

II. $\lim_{x \to 3^-} f(x) = \lim_{x \to 3^+} f(x) = f(3)$

III. $\lim_{x \to \infty} f(x) = 0$

(A) I only

(B) III only

(C) I and II only

(D) II and III only

11. Which of the following is the sixth-degree Taylor polynomial for e^{x^2} about $x = 0$?

(A) $1 + x + \dfrac{x^2}{2!} + \dfrac{x^3}{3!} + \dfrac{x^4}{4!} + \dfrac{x^5}{5!}$

(B) $1 + x^2 + \dfrac{x^4}{2!} + \dfrac{x^6}{3!}$

(C) $x - x^2 + \dfrac{x^4}{2!} - \dfrac{x^6}{2!}$

(D) $x - x^2 + \dfrac{x^4}{4!} - \dfrac{x^6}{6!}$

---

12. Let g be the function defined by $g(x) = \displaystyle\int_1^{\sqrt{x}} 2t + \frac{1}{t}\,dt$. Which of the following is the value of $g''(2)$?

(A) $\dfrac{11}{8}$ (B) $\dfrac{5}{4}$ (C) $\dfrac{3}{4}$ (D) $-\dfrac{1}{8}$

13. Let T be the third-degree Taylor polynomial for f about $x = 0$. If $f'(x) = (2x+4)^{\frac{3}{2}}$, which of the following is the coefficient of x^3 in T ?

(A) $\frac{2}{3}$ (B) $\frac{1}{2}$ (C) $\frac{1}{3}$ (D) $\frac{1}{4}$

---

14. Let $g(x) = f(1-x)$. If $f'(x) = (x-1)^2(3x+1)$, which of the following is the value of $g'(1)$?

(A) -3 (B) -1 (C) 2 (D) 4

15. If f is the function with $f'(x) = x\sqrt{x-1}$, at what value of x does f have an inflection point?

(A) $\frac{2}{3}$ (B) $\frac{1}{3}$ (C) $-\frac{1}{4}$ (D) nonexistent

16. Which of the following point (x, y) on the graph of $y = \sqrt{x}$ is closest to the point $(4, 0)$?

(A) $(3, \sqrt{3})$

(B) $\left(\frac{7}{2}, \frac{\sqrt{14}}{2}\right)$

(C) $(4, 2)$

(D) $\left(\frac{9}{2}, \frac{3\sqrt{2}}{2}\right)$

17. Which of the following is the right Riemann sum approximation of $\int_0^{\frac{\pi}{2}} \sin x\, dx$ using three subintervals of equal length?

(A) $\frac{\pi}{6}\left(\frac{1+\sqrt{3}}{2}\right)$

(B) $\frac{\pi}{6}\left(\frac{3+\sqrt{3}}{2}\right)$

(C) $\frac{\pi}{2}\left(\frac{\sqrt{3}-1}{2}\right)$

(D) $\frac{\pi}{2}\left(\frac{\sqrt{3}-3}{2}\right)$

---

18. The region R in the first quadrant is bounded by the graph of $y = \frac{1}{\sqrt{1-x^2}}$, the x-axis, the y-axis, and the line $x = k$. If the area of R is $\frac{\pi}{6}$, which of the following is the value of k ?

(A) $\frac{1}{2}$ (B) $\frac{\sqrt{2}}{2}$ (C) $\frac{\sqrt{3}}{2}$ (D) 1

19. Consider the equation $\ln y + xy = x^2$, where y is a function of x. Which of the following is the value of $\frac{dy}{dx}$ when $x = 1$?

(A) 2 (B) $\frac{3}{2}$ (C) $\frac{1}{2}$ (D) $\frac{1}{6}$

---

20. $\int_1^e \ln x \, dx =$

(A) $\frac{9}{4}$ (B) 2 (C) 1 (D) $\frac{1}{3}$

21. Which of the following infinite series converges?

I. $1 - 1 + 1 - 1 + \cdots + (-1)^{n+1} + \cdots$

II. $\frac{1}{2} + \frac{1}{3} + \frac{1}{4} + \cdots + \frac{1}{n+1} + \cdots$

III. $\cos 1 + \cos 2 + \cos 3 + \cdots + \cos n + \cdots$

(A) I only

(B) III only

(C) II and III only

(D) None

---

22. $\int \tan^2 x \sec^4 x \, dx =$

(A) $\frac{1}{3} \ln |\tan^2 x| + C$

(B) $\frac{1}{3} \ln |\sec x \tan x| + C$

(C) $\frac{1}{3} \tan^3 x + \frac{1}{5} \tan^5 x + C$

(D) $\frac{1}{3} \sec^3 x + \frac{1}{5} \sec^5 x + C$

23. Let $y = f(x)$ be the solution to the logistic differential equation $\dfrac{dy}{dx} = \dfrac{y}{5} - \dfrac{y^2}{2000}$. Which of the following value does y approach when x is large?

(A) 2000 (B) 800 (C) 400 (D) 100

---

24. Which of the following gives the area of the region that lies inside the polar graph $r = 4 + 4\sin\theta$ and lies outside the polar graph $r = 6$?

(A) $\dfrac{1}{2}\displaystyle\int_{\frac{\pi}{3}}^{\frac{2\pi}{3}} 6^2 - (4 + 4\sin\theta)^2\, d\theta$

(B) $\dfrac{1}{2}\displaystyle\int_{\frac{\pi}{3}}^{\frac{2\pi}{3}} (4 + 4\sin\theta)^2 - 6^2\, d\theta$

(C) $\dfrac{1}{2}\displaystyle\int_{\frac{\pi}{6}}^{\frac{5\pi}{6}} 6^2 - (4 + 4\sin\theta)^2\, d\theta$

(D) $\dfrac{1}{2}\displaystyle\int_{\frac{\pi}{6}}^{\frac{5\pi}{6}} (4 + 4\sin\theta)^2 - 6^2\, d\theta$

25. Which of the following is the coefficient of x^5 in the Maclaurin series for $x\cos x$?

(A) $\frac{1}{120}$ (B) $\frac{1}{24}$ (C) $-\frac{1}{120}$ (D) $-\frac{1}{24}$

---

26. What is the solution to the differential equation $\frac{dy}{dx} = x^2 y$ that satisfies $y(1) = e$?

(A) $y = e^{1/2(x^2+1)}$

(B) $y = e^{1/2(-x^2+3)}$

(C) $y = e^{1/3(x^3+2)}$

(D) $y = e^{1/3(-x^3+4)}$

27. Which of the following is the interval of convergence of the power series $\sum_{n=1}^{\infty} \frac{(-2)^n(2x-1)^n}{\sqrt{n}}$?

(A) $\frac{1}{4} < x \le \frac{3}{4}$

(B) $\frac{1}{4} \le x \le \frac{3}{4}$

(C) $\frac{1}{2} \le x < 1$

(D) $\frac{1}{2} < x < 1$

---

x	f	g
0	5	2
2	7	4

28. The values of the functions f and g at $x = 0$ and $x = 2$ are shown in the table above. If $\int_0^2 f'(x)g(x)\,dx = 7$, which of the following is the value of $\int_0^2 f(x)g'(x)\,dx$?

(A) 5 (B) 8 (C) 9 (D) 11

29. A particle moves in the xy-plane so that its position vector is $< t^3 - 6t^2 + 9t, t^3 - \frac{15}{2}t^2 + 18t >$. At which of the following value of t does the particle change its direction?

(A) $t = 2$ only

(B) $t = 3$ only

(C) $t = 1$ and $t = 3$

(D) $t = 2$ and $t = 3$

---

30. Let f be the function defined by $f(x) = \int_{-2}^{x} \sqrt{4 - x^2}\, dx$. Which of the following is the slope of the secant line that passes through the points $(-2, f(-2))$ and $(2, f(2))$?

(A) $\frac{\pi}{2}$ (B) $\frac{3\pi}{4}$ (C) π (D) $\frac{5\pi}{4}$

END OF PART A OF SECTION I

STOP

CALCULUS BC TEST 4
SECTION I, Part B
Time — 45 minutes
Number of questions — 15

A GRAPHING CALCULATOR IS REQUIRED FOR SOME QUESTIONS ON THIS PART OF THE EXAM.

Directions: Solve each of the following problems using the available space for scratch work. Choose the best answer among the answer choices given and fill in the corresponding circle on the answer sheet.

76. Which of the following is the area of the region in the first quadrant bounded by the graphs of $y = \dfrac{1}{\sqrt{1+x}}$ and $y = \dfrac{1}{2}x^2$?

(A) 0.313 (B) 0.572 (C) 0.679 (D) 0.809

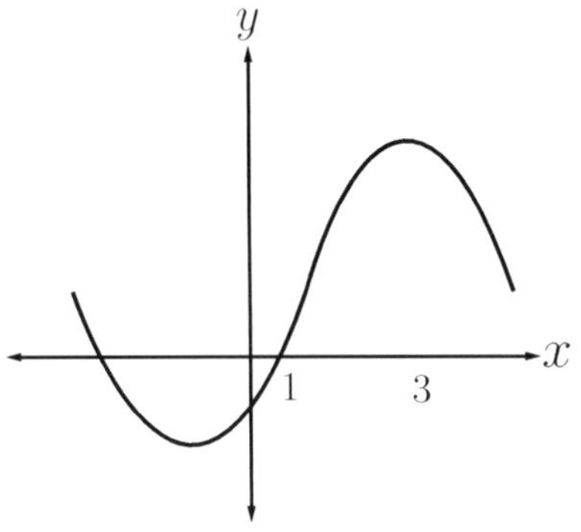

Graph of f

77. The graph of a differentiable function f is shown above. If $g(x) = 2x^3 - 3x^2 - 12x + 24$ and the function h is defined by $h(x) = f(x) \cdot g(x)$, which of the following must be true about $h'(2)$?

(A) $h'(2) > 0$

(B) $h'(2) = 0$

(C) $h'(2) < 0$

(D) Undefined

---

78. The region R is enclosed by the graph of $y = \sin x$, the x-axis, and two lines $x = 0$ and $x = \dfrac{\pi}{2}$. What is the volume of a solid if R is rotated about the line $x = -1$?

(A) 6.283 (B) 7.248 (C) 12.566 (D) 14.495

x	1	2	4	6	7	11	12
$f(x)$	2	6	3	4	8	3	1

79. Let f be a differentiable function on the closed interval $[1, 12]$. The table above shows values of f for selected values of x. All of the following statements are true EXCEPT

(A) There is a number c in the open interval $(2, 6)$ such that $f(c) = 4.5$.

(B) There is a number c in the open interval $(7, 11)$ such that $f(c) = 6$.

(C) There is a number c in the open interval $(4, 11)$ such that $f'(c) = 0$.

(D) There is a number c in the open interval $(1, 7)$ such that $f'(c) = -1$.

---

80. If $\int_1^5 f(x)\,dx = 6$, which of the following is the value of $\int_0^2 f(2x+1)\,dx$?

(A) 2 (B) 3 (C) 6 (D) 12

81. The velocity a particle moving in the xy-plane are given by the parametric equation $x'(t) = 1 + t^2$ and $y'(t) = te^t$. At what time t is the tangent line to the path of the particle is parallel to the line $y = 3x + 4$?

(A) $t = 1.813$

(B) $t = 2.021$

(C) $t = 2.549$

(D) $t = 2.974$

---

82. The surface area of a cube is increasing at a rate of 6 cm^2/min. How fast is the volume of the cube is increasing when the surface area of the cube is 9 cm^2 ?

(A) $\dfrac{4}{\sqrt{6}}$ cm^3/min

(B) $\dfrac{4}{3\sqrt{6}}$ cm^3/min

(C) $\dfrac{4\sqrt{6}}{9}$ cm^3/min

(D) $\dfrac{3\sqrt{6}}{4}$ cm^3/min

83. The population of bears in a forest at $t = 0$ year is 150. The rate at which the population of bears grows is proportional to its size. If the population of bears at $t = 2$ years is 302, what is the population of bears at $t = 6$ years?

(A) 624 (B) 817 (C) 1225 (D) 1533

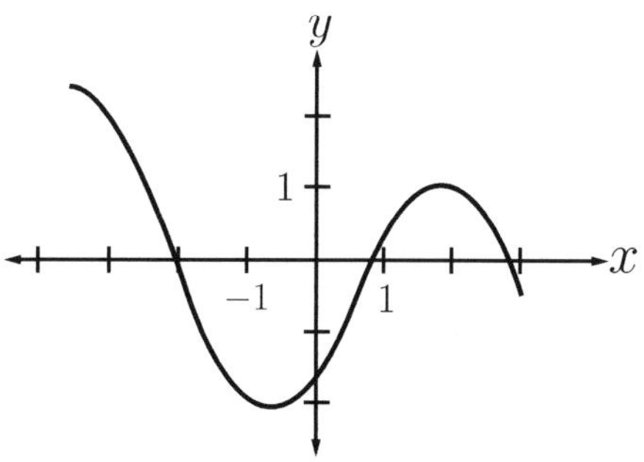

Graph of f

84. The graph of the function f for $-3.5 \le x \le 3$ is shown in the figure above. If the function g is defined by $g(x) = \int_1^x f(t)\,dt$, for which of the following value of x does g attain the absolute maximum?

(A) -3 (B) -2 (C) 1 (D) 2

85. Which of the following statement must be true?

(A) If $\lim_{n\to\infty} \dfrac{a_n}{b_n} = 1$ and $\sum_{n=1}^{\infty} a_n$ converges, then $\sum_{n=1}^{\infty} b_n$ converges.

(B) If $\lim_{n\to\infty} a_n = 0$, then $\sum_{n=1}^{\infty} a_n$ converges.

(C) If $b_n < a_n$ and $\sum_{n=1}^{\infty} a_n$ diverges, then $\sum_{n=1}^{\infty} b_n$ converges.

(D) If $\sum_{n=1}^{\infty} (-1)^n a_n$ converges, then $\sum_{n=1}^{\infty} \left|(-1)^n a_n\right|$ converges.

---

86. The function f is given by $f(x) = \int_{-1}^{x^2} \sqrt{e^t + 1}\,dt$. Which of the following is the value of c that satisfies the conclusion of the Mean Value Theorem on the closed interval $[0, 1]$?

(A) 0.866 (B) 0.679 (C) 0.450 (D) 0.372

87. A particle moves along the x-axis so that its velocity is given by $v(t) = -\sin t$ for $0 \le t \le 3$. At $t = 0$, the particle is located at 1 unit to the left of the origin. Which of the following statement must be true?

(A) The particle passes through the origin for some time t between $t = 0$ and $t = 3$.

(B) The particle moves to the left and then moves to the right from $t = 0$ to $t = 3$.

(C) The particle is at rest for some time t between $t = 1$ and $t = 2$.

(D) The particle is located at 2.990 units to the left of the origin at $t = 3$.

$$T(x) = \sum_{n=0}^{\infty} \frac{(-1)^n x^n}{n!} = 1 - x + \frac{x^2}{2!} - \frac{x^3}{3!} + \cdots$$

88. The function f has derivatives of all orders, and the Maclaurin series T for f is given above. If the approximation for $f\left(\frac{1}{2}\right)$ is obtained by using the first three terms of the series T, which of the following is the maximum error for this approximation?

(A) $\frac{1}{48}$ (B) $\frac{1}{96}$ (C) $\frac{1}{120}$ (D) $\frac{1}{192}$

89. If the function $f(x)$ is defined by $f(x) = \sqrt{x^3+1}$, which of the following is the average value of f over the interval $[1, 3]$?

(A) 2.923 (B) 3.115 (C) 3.587 (D) 3.938

---

90. A particle moves in the xy-plane so that its position are given by the parametric equations $x(t) = te^{-t}$ and $y(t) = \dfrac{1}{2t^2+1}$. Which of the following is the speed of the particle at $t = 2$?

(A) 0.731 (B) 0.405 (C) 0.238 (D) 0.168

END OF PART B OF SECTION I

STOP

CALCULUS BC TEST 5

SECTION I, Part A

Time — 60 minutes

Number of questions — 30

A CALCULATOR MAY NOT BE USED ON THIS PART OF THE EXAM.

Directions: Solve each of the following problems using the available space for scratch work. Choose the best answer among the answer choices given and fill in the corresponding circle on the answer sheet.

1. If $y = e^{-2x}$, then $\dfrac{d^2y}{dx^2} =$

   (A) $2e^{2x}$

   (B) $2e^{-2x}$

   (C) $4e^{2x}$

   (D) $4e^{-2x}$

2. If the Maclaurin series for the function f is given by $f(x) = \sum_{n=0}^{\infty} \frac{(-1)^n}{3^n} x^n$, which of the following is the value of $f(2)$?

(A) 3 (B) $\frac{7}{3}$ (C) $\frac{3}{5}$ (D) $\frac{2}{3}$

---

3. A particle moves along the curve $y = \frac{2}{3}\sqrt{x^3}$. If the particle moves from the point $(0, 0)$ to the point $(9, 18)$, which of the following gives the total distance traveled by the particle?

(A) $\int_0^{18} 1 + \sqrt{x}\, dx$

(B) $\int_0^{18} \sqrt{1 + \sqrt{x}}\, dx$

(C) $\int_0^{9} \sqrt{1 + x}\, dx$

(D) $\int_0^{9} 1 + x\, dx$

4. If the function f is defined by $f(x) = \dfrac{|x-2|}{x-2}$, what is the value of $\displaystyle\int_0^3 f(x)\,dx$?

(A) -1 (B) 0 (C) 1 (D) 2

5. The position of a particle moving in the xy-plane is given by the parametric equations $x(t) = t^2 - 4t$ and $y(t) = 3t - t^3$. At what time t does the particle move to the right and downward?

(A) $t = 3$ (B) $t = 2$ (C) $t = 1$ (D) $t = 0$

6. Which of the following series converges?

I. $\sum_{n=0}^{\infty} e^{-n}$ II. $\sum_{n=0}^{\infty} \frac{1}{\sqrt{n^3+1}}$ III. $\sum_{n=0}^{\infty} (-1)^n \frac{n!}{2^n}$

(A) I only

(B) II only

(C) I and II only

(D) II and III only

---

7. $\int_0^2 x\sqrt{1+2x^2}\,dx =$

(A) $\frac{42}{3}$ (B) $\frac{26}{3}$ (C) $\frac{19}{3}$ (D) $\frac{13}{3}$

8. If the function f is given by $f(x) = \dfrac{1}{x}$, which of the following best describes $f^{(n)}(x)$, the nth order derivative of f ?

(A) $f^{(n)}(x) = \dfrac{n!}{x^n}$

(B) $f^{(n)}(x) = \dfrac{n!}{x^{n+1}}$

(C) $f^{(n)}(x) = (-1)^n \dfrac{n!}{x^n}$

(D) $f^{(n)}(x) = (-1)^n \dfrac{n!}{x^{n+1}}$

---

9. If $\dfrac{1}{y} = e^{\arctan x}$, then $\dfrac{dy}{dx} =$

(A) $-\dfrac{xye^{\arctan x}}{1+x^2}$

(B) $-\dfrac{y^2e^{\arctan x}}{1+x^2}$

(C) $-\dfrac{xe^{\arctan x}}{y}$

(D) $-\dfrac{e^{\arctan x}}{y}$

t	3	4	6	9
$R(t)$	6	4	5	9

10. The rate at which water flows into a tank is $R(t)$, where $R(t)$ is measured in cubic feet per minute, t is measured in minutes, and $3 \le t \le 9$. The tank contains 23 cubic feet of water at $t = 3$ minutes. The table above shows values of $R(t)$ for selected values of t. What is the approximation of the amount of water, in cubic feet, in the tank at $t = 9$ minutes using a trapezoidal sum with three subintervals of unequal length?

(A) 51 (B) 58 (C) 64 (D) 72

---

11. If the function g is defined by $g(x) = \lim_{h \to 0} \dfrac{\cos(x+h) - \cos x}{h}$, which of the following must be true?

(A) $g\left(\frac{\pi}{2}\right) < g''\left(\frac{\pi}{2}\right) < g'\left(\frac{\pi}{2}\right)$

(B) $g\left(\frac{\pi}{2}\right) < g'\left(\frac{\pi}{2}\right) < g''\left(\frac{\pi}{2}\right)$

(C) $g'\left(\frac{\pi}{2}\right) < g\left(\frac{\pi}{2}\right) < g''\left(\frac{\pi}{2}\right)$

(D) $g'\left(\frac{\pi}{2}\right) < g''\left(\frac{\pi}{2}\right) < g\left(\frac{\pi}{2}\right)$

12. Let f be the even function so that $f(-x) = f(x)$ for all real numbers x. If $f'(c)$ exists, which of the following must be $f'(-c)$ in terms of $f'(c)$?

(A) $\dfrac{1}{f'(c)}$ (B) $-\dfrac{1}{f'(c)}$ (C) $f'(c)$ (D) $-f'(c)$

---

13. The acceleration $a(t)$ of a particle moving in the y-axis is given by $a(t) = \dfrac{3}{2}\sqrt{t} - 1$. At $t = 1$, the velocity of the particle is 2. If $s(t)$ is the position of the particle from the origin at time t, which of the following is the value of $s(1) - s(0)$?

(A) $\dfrac{23}{7}$ (B) $\dfrac{19}{10}$ (C) $\dfrac{3}{2}$ (D) $\dfrac{4}{5}$

14. Let $y = f(x)$ be the solution to the differential equation $\dfrac{dy}{dx} = 4x - y$. Which of the following must be true?

(A) The point $(1, 1)$ is a critical number of f.

(B) f is concave up at the point $(2, 3)$.

(C) The point $(1, 4)$ is a relative minimum of f.

(D) f has an inflection point at $(2, 5)$.

---

15. Which of the following is the radius of convergence of the series $\sum_{n=0}^{\infty}(-1)^n \dfrac{(x-3)^n}{n!}$?

(A) 0 (B) $\dfrac{1}{2}$ (C) 3 (D) ∞

16. If the function f is defined by $f(x) = \dfrac{\ln x}{x}$ for $x > 0$, on which of the following interval is f increasing?

(A) $(0, 1)$

(B) $(0, e)$

(C) $(1, e)$

(D) (e, ∞)

---

17. Which of the following is the Maclaurin series for $e^x + e^{-x}$?

(A) $\dfrac{x^2}{2 \cdot 2!} + \dfrac{x^4}{2 \cdot 4!} + \dfrac{x^6}{2 \cdot 6!} \cdots$

(B) $\dfrac{x^2}{2!} + \dfrac{x^4}{4!} + \dfrac{x^6}{6!} \cdots$

(C) $2 + \dfrac{x^2}{2!} + \dfrac{x^4}{4!} + \cdots$

(D) $2 + \dfrac{2x^2}{2!} + \dfrac{2x^4}{4!} + \cdots$

18. $\displaystyle\int_0^1 \frac{x+5}{x^2+4x+3}\,dx =$

(A) $\ln 3$ (B) $2\ln 5$ (C) $2\ln 7$ (D) $\ln 56$

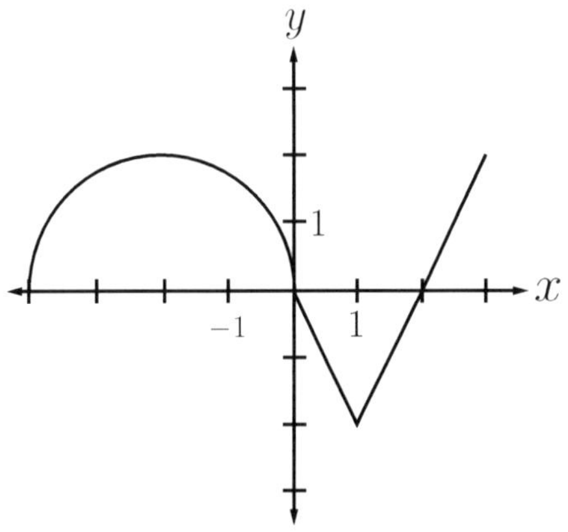

Graph of f'

19. The graph of f', the derivative of f, shown above consists of a semicircle and two line segments. If $f(3) = 5$, then which of the following is the value of $f(-2)$?

(A) $6 - \pi$ (B) $4 - \pi$ (C) $\pi - 3$ (D) $\pi - 2$

20. Let $y = f(x)$ be the solution to the differential equation $\dfrac{dy}{dx} = x^2 - y$ with initial condition $f(2) = 1$. Which of the following is the approximation for $f(3)$, starting at $x = 2$, obtained by using Euler's method with two steps of equal length?

(A) 3.225 (B) 3.750 (C) 4.375 (D) 4.725

---

21. $\displaystyle\int_{-\infty}^{1} \frac{1}{\sqrt{3-x}}\, dx =$

(A) $\dfrac{1}{2}$ (B) $\sqrt{2}$ (C) $2\sqrt{3}$ (D) Divergent

22. $\displaystyle\int_0^2 |x^2 - 1|\, dx =$

(A) $\dfrac{1}{3}$ (B) 1 (C) $\dfrac{4}{3}$ (D) 2

---

23. Let f be a differentiable function. If $\displaystyle\int f(x)\, dx = xf(x) - \int 1\, dx$, which of the following must be $f(x)$?

(A) $\ln x$ (B) $\tan^{-1} x$ (C) xe^x (D) e^{-x^2}

24. If the power series $\sum_{n=0}^{\infty} C_n(x-1)^n$ converges at $x=4$, which of the following could be C_n ?

(A) $(-1)^n \dfrac{1}{4^n}$

(B) $(-1)^n \dfrac{1}{3^n}$

(C) $(-1)^n \dfrac{1}{2^n}$

(D) $\dfrac{1}{2^n}$

---

25. The region R is enclosed by the graph of $y=e^{-x^2}$, the y-axis, the x-axis, and the line $x=2$. Which of the following is the volume of the solid if R is rotated about the y-axis?

(A) $\pi\left(\dfrac{e^2-e}{e^2}\right)$

(B) $\pi\left(\dfrac{e^4-1}{e^4}\right)$

(C) $2\pi\left(\dfrac{e^2-e}{e^2}\right)$

(D) $2\pi\left(\dfrac{e^4-1}{e^4}\right)$

26. If $y = x^{x \ln x}$, then $\dfrac{dy}{dx} =$

(A) $x^{1/x}$

(B) $x^{x \ln x} \cdot \ln x$

(C) $x^{x \ln x}(\ln x + 1)$

(D) $x^{x \ln x}\left((\ln x)^2 + 2 \ln x\right)$

---

27. Which of the following could be the value of θ for which the polar curve $r = 1 - \sin\theta$ for $0 \le \theta \le \pi$ has a vertical tangent?

(A) $\dfrac{\pi}{2}$ (B) $\dfrac{\pi}{3}$ (C) $\dfrac{\pi}{4}$ (D) $\dfrac{\pi}{6}$

28. For $t \geq 0$, the position of a particle moving in the xy-plane is given by $x(t) = \sqrt{5-t}$ and $y(t) = \dfrac{1}{3-t}$. which of the following is the acceleration vector of the particle at $t = 4$?

(A) $\left\langle -\frac{1}{4}, -2 \right\rangle$ (B) $\left\langle -\frac{1}{2}, -1 \right\rangle$ (C) $\left\langle \frac{1}{2}, -\frac{1}{2} \right\rangle$ (D) $\left\langle \frac{1}{4}, -1 \right\rangle$

29. Let f be a differentiable function with $f(3) = -2$ and $f(6) = 4$. Which of the following is the value of $\lim\limits_{x \to 3} \dfrac{\ln(4-x)}{\displaystyle\int_{6}^{2x} f(t)\,dt}$?

(A) $\frac{1}{2}$ (B) $\frac{1}{4}$ (C) $-\frac{1}{8}$ (D) $-\frac{1}{4}$

30. Let f be a differentiable function such that $f(x) > 0$ for all x and $f(-x) = f(x)$ for all x. All of the following statements are true EXCEPT

(A) $\int_b^c f(x)\,dx = \int_a^c f(x)\,dx - \int_a^b f(x)\,dx$

(B) $\int_{-a}^a f(x)\,dx = 2\int_{-a}^0 f(x)\,dx$

(C) $\int_{-a}^a f(x)\,dx = -\int_a^{-a} f(x)\,dx$

(D) $\int_{-a}^a f(x)\,dx = 0$

END OF PART A OF SECTION I

STOP

CALCULUS BC TEST 5
SECTION I, Part B
Time — 45 minutes
Number of questions — 15

A GRAPHING CALCULATOR IS REQUIRED FOR SOME QUESTIONS ON THIS PART OF THE EXAM.

Directions: Solve each of the following problems using the available space for scratch work. Choose the best answer among the answer choices given and fill in the corresponding circle on the answer sheet.

76. The derivative of f, f', is given by $f'(x) = 2(x+1)(x-2)(x-4)$. Which of the following statements must be true?

I. f has a relative minimum at $x = -1$ and $x = 4$.

II. f is increasing in the interval $(2, 4)$.

III. f is concave down in the interval $(0.214, 3.120)$.

(A) I only

(B) II only

(C) I and III only

(D) II and III only

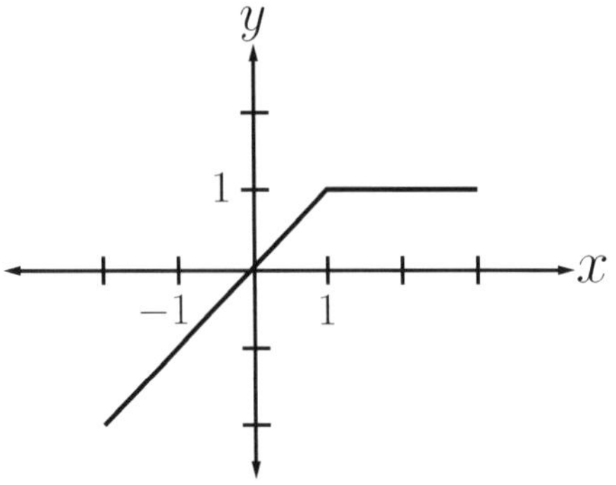

Graph of f

77. The graph of f in the figure above consists of two line segments. Which of the following is the value of $\int_{-2}^{3} 2f(x) - 1\,dx$?

(A) 5 (B) 2 (C) −4 (D) −6

---

78. If the function f is defined by $f(x) = \sin x$, which of the following is the fourth-degree Taylor series for f about $x = \frac{\pi}{2}$?

(A) $1 - \left(x - \frac{\pi}{2}\right)^2 + \frac{1}{2!}\left(x - \frac{\pi}{2}\right)^4$

(B) $1 - \frac{1}{2!}\left(x - \frac{\pi}{2}\right)^2 + \frac{1}{4!}\left(x - \frac{\pi}{2}\right)^4$

(C) $x - \frac{1}{3!}\left(x - \frac{\pi}{2}\right)^3 + \frac{1}{5!}\left(x - \frac{\pi}{2}\right)^5$

(D) $x - \frac{1}{3!}\left(x - \frac{\pi}{2}\right)^3 + \frac{1}{5!}\left(x - \frac{\pi}{2}\right)^5 - \frac{1}{7!}\left(x - \frac{\pi}{2}\right)^7$

79. For which of the following value of x does the slope of the tangent line to the graph of $y = \ln x$ equal the slope of the tangent line to the graph of $y = e^x$?

(A) 0.391 (B) 0.421 (C) 0.452 (D) 0.567

$$f(x) = \begin{cases} \sqrt{2x+5}, & -2.5 \le x < 2 \\ 7 - 2x, & 2 \le x \le 5 \end{cases}$$

80. If the piecewise function f is defined above, which of the following must be true?

(A) f is discontinuous at $x = 2$.

(B) f is not differentiable at $x = 2$.

(C) f is increasing on the closed interval $[-2.5, 5]$.

(D) f is concave up on the open interval $(-2.5, 2)$.

Car	Rate of fuel consumption
A	$R(t) = 10 + 5\cos(t-1)$
B	$L(t) = 5 + e^{0.5t}$

81. The rates of fuel consumption of car A and car B, in gallons per hour, where t is measured in hours, are shown in the table above. If both cars travel on a highway at the same time from $t = 0$ hour to $t = 6$ hours, which of the following statement must be true?

(A) Car A consumes 8.758 gallons of fuel more than car B does.

(B) Car A consumes 7.955 gallons of fuel more than car B does.

(C) Car B consumes 8.758 gallons of fuel more than car A does.

(D) Car B consumes 7.955 gallons of fuel more than car A does.

---

82. Let f be a differentiable function on the closed interval $[-2, 3]$ with $f(-1) = f(2)$. Which of the following must be true?

(A) There must be c on the interval $(-1, 2)$ such that $f'(c) = 0$.

(B) There must be c on the interval $(-2, -1)$ such that $f''(c) = 0$.

(C) f has a local minimum on the open interval $(-2, 3)$.

(D) $f'(-1) = f'(2)$

	$x < 2$	$2 < x < 4$	$4 < x < 5$	$x > 5$
g'	Positive	Negative	Positive	Negative
g''	Negative	Negative	Positive	Positive

83. The table above shows the signs of g' and g'', the first and the second derivatives of g. Which of the following interval is g decreasing and concave up?

(A) $x < 2$ (B) $2 < x < 4$ (C) $4 < x < 5$ (D) $x > 5$

---

84. Consider a curve defined by the parametric equations $x(t) = \frac{1}{2}t^2 - 3t$ and $y(t) = t^3 - 3t^2 - 9t$. Which of the following is the value of $\frac{d^2y}{dx^2}$ at $t = 2$?

(A) -3 (B) -2 (C) $-\frac{1}{2}$ (D) $-\frac{1}{3}$

85. Which of the following is the average value of $y = \sqrt[3]{x^4 + 1}$ on the interval $[0, 4]$?

(A) 2.191 (B) 2.556 (C) 2.624 (D) 2.920

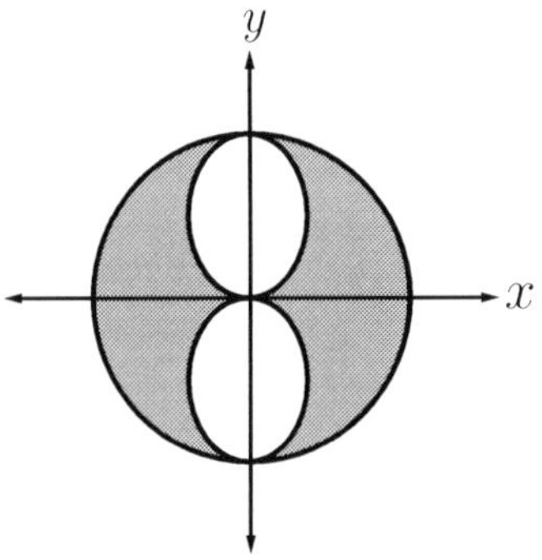

86. The figure above shows the graphs of $r = 2 - 2\cos 2\theta$ and $r = 4$. which of the following is the area of the shaded region?

(A) 28.832 (B) 29.651 (C) 31.416 (D) 33.061

87. A particle moves in the xy-plane so that its velocity is given by $v(t) = \sin(t^2)$ for $t \geq 0$. At $t = 1$, the particle is located at 0.296 unit right of the origin. Which of the following best represents the displacement of the particle at $t = 3$?

(A) The particle is located at 0.759 unit right of the origin.

(B) The particle is located at 0.577 unit right of the origin.

(C) The particle is located at 0.284 unit left of the origin.

(D) The particle is located at 0.367 unit left of the origin.

---

88. Consider the series $\sum_{n=1}^{\infty} a_n$. Which of the following could be a_n for which the series diverges?

(A) $a_n = (-1)^n \dfrac{1}{3n+1}$

(B) $a_n = \dfrac{1}{n\sqrt{n+1}}$

(C) $a_n = \left(1 + \dfrac{1}{n}\right)^n$

(D) $a_n = \dfrac{3^n + 4^n}{7^n}$

89. Which of the following statement must be true?

(A) If f is continuous at $x = a$, then $f'(a)$ exists.

(B) If $\lim_{x \to a^+} f(x)$ exists and $\lim_{x \to a^-} f(x)$ exits, then $\lim_{x \to a} f(x)$ exits.

(C) If $\lim_{x \to 2} \frac{f(x) - f(2)}{x - 2} = 5$, then $\lim_{h \to 0} \frac{f(2 + h) - f(2)}{h} = 5$.

(D) If $f(x) = x^n$, then $f^{(n)}(x) = (n + 1)!$.

---

90. The derivative of f, f', is given by $f'(x) = x^2 \sin(x^2)$. Which of the following is the number of inflection points does the graph of f have on the open interval $(-2, 2)$?

(A) None (B) 1 (C) 2 (D) 3

END OF PART B OF SECTION I

STOP

CALCULUS BC TEST 6
SECTION I, Part A
Time — 60 minutes
Number of questions — 30

A CALCULATOR MAY NOT BE USED ON THIS PART OF THE EXAM.

Directions: Solve each of the following problems using the available space for scratch work. Choose the best answer among the answer choices given and fill in the corresponding circle on the answer sheet.

1. If h approaches 0, which of the following best approximates the value of $\dfrac{(2+h)^2 - 2^2}{h}$?

(A) 1 (B) 2 (C) 3 (D) 4

2. Which of the following is the equation of the tangent line to the graph of $y = \ln(e - x)$ at the point $(e - 1, 0)$?

(A) $y = x - e + 1$

(B) $y = x + e - 1$

(C) $y = -x - e + 1$

(D) $y = -x + e - 1$

---

3. If $y = \dfrac{\sin x - 1}{\cos x}$, then $\dfrac{dy}{dx} =$

(A) $\dfrac{\cos x - 1}{\cos^2 x}$

(B) $\dfrac{\cos x + 1}{\cos^2 x}$

(C) $\dfrac{1 + \sin x}{\cos^2 x}$

(D) $\dfrac{1 - \sin x}{\cos^2 x}$

4. Let S_n be the partial sum defined by $S_n = \sum_{k=0}^{n} (-1)^k \left(\frac{2}{5}\right)^k$. Which of the following is the value of $\lim_{n\to\infty} S_n$?

(A) $\frac{5}{3}$ (B) $\frac{5}{7}$ (C) $\frac{2}{5}$ (D) $-\frac{2}{7}$

---

5. The acceleration of a particle moving in the y-axis at any time $t \geq 0$ is given by $a(t) = \frac{1}{1+t^2}$. If the velocity of the particle at time $t = 0$ is $v(0) = \frac{\pi}{2}$, which of the following is $v(1)$?

(A) π (B) $\frac{3\pi}{4}$ (C) $\frac{\pi}{3}$ (D) $-\frac{\pi}{4}$

6. $\lim_{x \to \infty} (xe^{1/x} - x) =$

(A) 1 (B) $\frac{1}{2}$ (C) $\frac{1}{e}$ (D) nonexistent

---

7. Let T be the Maclaurin series for the function f defined by $f(x) = e^{x^2} + \cos x$. Which of the following is the coefficient of x^4 in T ?

(A) $\frac{7}{24}$ (B) $\frac{11}{24}$ (C) $\frac{13}{24}$ (D) $\frac{17}{24}$

8. $\displaystyle\int \frac{x+2}{\sqrt{x-1}}\,dx =$

(A) $\frac{2}{3}\sqrt{x-1} + 6\frac{1}{\sqrt{x-1}} + C$

(B) $\frac{2}{3}\sqrt{x-1} - 6\frac{1}{\sqrt{x-1}} + C$

(C) $\frac{2}{3}(x-1)\sqrt{x-1} + 6\sqrt{x-1} + C$

(D) $\frac{2}{3}(x-1)\sqrt{x-1} - 6\sqrt{x-1} + C$

---

x	1	2	3	4	5	6	7
$f(x)$	$n-1$	n	$n+1$	$n-2$	$n-3$	$n+4$	$2n$

9. Let f be a continuous function on the closed interval $[1, 7]$. The table above shows values of f for selected values of x in terms of n. If the approximation for $\int_1^7 f(x)\,dx$ obtained by a midpoint sum with three subintervals with equal length is 40, then which of the following is the value of n ?

(A) 8 (B) 6 (C) 5 (D) 4

10. Let $y = f(x)$ be the solution to the differentiable equation $\frac{dy}{dx} = y^2 - x^2$. At which of the following point (x, y) on the graph of f does f has a local maximum?

(A) $(-2, 2)$ (B) $(-1, 3)$ (C) $(-1, 1)$ (D) $(2, -2)$

---

x	-1	0	2
f	2	-1	3
f'	4	2	1

11. The table above shows values of f and its first derivative, f' for selected values of x. Which of the following is the slope of the tangent line to the graph of the inverse function of f at $x = 2$?

(A) $\frac{1}{3}$ (B) $\frac{1}{4}$ (C) $-\frac{1}{2}$ (D) -1

12. $\displaystyle\int \frac{x}{1+x^4}\,dx =$

(A) $2\tan^{-1}\left(\frac{x^2}{2}\right)+C$

(B) $\frac{1}{2}\tan^{-1}(x^2)+C$

(C) $\ln|1+x^4|+C$

(D) $2\ln\left|\frac{x^2-1}{x^2+1}\right|+C$

---

13. The area of an equilateral triangle is increasing at a rate of $2\sqrt{3}$ cm^2/sec. Which of the following is the rate at which each side of the equilateral triangle is increasing when the side length of the equilateral triangle is 3 cm?

(A) $\frac{4}{3}$ cm/sec (B) $\frac{2}{3}$ cm/sec (C) $\frac{3}{5}$ cm/sec (D) $\frac{1}{4}$ cm/sec

14. If $f'(x) = -2xe^{-x^2}$ and $f(1) = \frac{3}{e}$, then what is the value of $f(0)$?

(A) $\frac{e}{e-2}$ (B) $\frac{e}{e-1}$ (C) $\frac{e+1}{e}$ (D) $\frac{e+2}{e}$

---

15. Which of the following is the sum of the series $1 - 2\ln 2 + \frac{(2\ln 2)^2}{2!} - \frac{(2\ln 2)^3}{3!} + \cdots$?

(A) e (B) $\frac{1}{2}$ (C) $\frac{1}{4}$ (D) $\frac{1}{e^2}$

16. Let f be a continuous, positive, and decreasing function on the interval $[1, \infty]$ such that $a_n = f(n)$. If $\sum_{n=1}^{\infty} a_n = 3$, which of the following must be true?

(A) $\int_1^{\infty} f(x)\, dx$ diverges

(B) $\int_1^{\infty} f(x)\, dx$ converges

(C) $\int_1^{\infty} f(x)\, dx = 3$

(D) $\lim_{n\to\infty} a_n = 3$

---

17. Let R be the region between the graph of $y = e^{-0.2x}$ and the x-axis for $x \geq 5$. Which of the following is the area of R ?

(A) $2e$ (B) $\frac{5}{e}$ (C) $\frac{1}{2e}$ (D) $\frac{1}{5e}$

18. Which of the following series converges for all real numbers x ?

I. $\sum_{n=1}^{\infty} \frac{x^n}{\sqrt{n}}$ II. $\sum_{n=1}^{\infty} \frac{x^n}{n^2}$ III. $\sum_{n=1}^{\infty} \frac{x^n}{n!}$

(A) III only

(B) I and II only

(C) I and III only

(D) II and III only

---

19. Which of the following is the slope of the tangent line to the graph of $x^2 + xy - y^2 = -1$ at the point $(1, 2)$?

(A) $-\frac{4}{5}$ (B) $-\frac{3}{4}$ (C) $\frac{4}{3}$ (D) $\frac{5}{4}$

20. If $f(x) = \arccos(2x)$, then $f'(x) =$

(A) $\dfrac{-2}{\sqrt{1-x^4}}$

(B) $\dfrac{2}{\sqrt{1-x^4}}$

(C) $\dfrac{-2}{\sqrt{1-4x^2}}$

(D) $\dfrac{2}{\sqrt{1-4x^2}}$

---

21. Which of the following gives the length of the path defined by the parametric equations $x(t) = \cos(t^2)$ and $y(t) = e^{-2t}$ from $t = 0$ to $t = 2\pi$?

(A) $\displaystyle\int_0^{2\pi} \sqrt{\cos^2(t^2) + e^{-4t}}\, dt$

(B) $\displaystyle\int_0^{2\pi} \sqrt{\cos^2(t^4) + 4e^{-4t}}\, dt$

(C) $\displaystyle\int_0^{2\pi} \sqrt{2t\sin^2(t^4) + 4e^{-4t}}\, dt$

(D) $\displaystyle\int_0^{2\pi} \sqrt{4t^2\sin^2(t^2) + 4e^{-4t}}\, dt$

22. Let R_4 and L_4 be a right Riemann sum and a left Riemann sum for $\int_0^1 x^2\,dx$ with 4 subintervals of equal length, respectively. Which of the following is the value of $R_4 - L_4$?

(A) $\frac{1}{8}$ (B) $\frac{3}{16}$ (C) $\frac{1}{4}$ (D) $\frac{5}{16}$

---

23. If the function f is defined by $f(x) = \ln(x-1) - \frac{1}{2}x + 4$, for which of the following value of x does f attain a local maximum?

(A) $\frac{4}{3}$ (B) $\frac{5}{2}$ (C) 3 (D) 5

24. Let f be a continuous function on the closed interval $[-3, 1]$ with $f(-3) = 2$ and $f(1) = 3$. If there is no c in the open interval $(-3, 1)$ such that $f'(c) = \frac{1}{4}$, which of the following must be true?

(A) There is no zero of f on the open interval $(-3, 1)$.

(B) $f'(x)$ does not exist for some x on the open interval $(-3, 1)$.

(C) $f'(x)$ exists for all real numbers x on the open interval $(-3, 1)$.

(D) $\lim_{x \to \infty} f(x) = 0$.

---

25. Which of the following is the particular solution to the differential equation $\frac{dy}{dx} = \frac{xy}{\ln y}$ with initial condition $y(0) = e^2$?

(A) $2 \ln y = \frac{1}{2}x^2 + 4$

(B) $2 \ln y = x^2 + 4$

(C) $(\ln y)^2 = \frac{1}{2}x^2 + 4$

(D) $(\ln y)^2 = x^2 + 4$

26. Which of the following is the radius of convergence of the series $\sum_{n=0}^{\infty} n!(x-2)^n$?

(A) 0 (B) $\frac{1}{2}$ (C) 2 (D) ∞

---

27. Let f be the function defined by $f(x) = \cos\left(\frac{1}{x^2}\right)$. Which of the following statements are true?

I. f is discontinuous at $x = 0$.

II. f is differentiable on the open interval $\left(-\frac{\pi}{2}, \frac{\pi}{2}\right)$.

III. The graph of f has a horizontal asymptote at $y = 1$.

(A) III only

(B) I and II only

(C) I and III only

(D) II and III only

28. If the function g is defined by $g(x) = \int_1^x t\sqrt{10 - t^2}\,dt$, where $x > 1$, for which of the following value of x does the graph of g change its concavity?

(A) $\sqrt{5}$ (B) 2 (C) $\sqrt{3}$ (D) $\sqrt{2}$

---

29. $\int_0^{\frac{\pi}{3}} x \sec^2 x\,dx =$

(A) $\dfrac{\pi}{\sqrt{3}} - \ln 2$ (B) $\dfrac{\pi}{3} - \ln 2$ (C) $\dfrac{\pi}{3\sqrt{3}} - \ln 2$ (D) $\dfrac{\sqrt{3}}{3}\pi - \ln 2$

30. Which of the following values of x does the series $\sum_{n=1}^{\infty}\left(\frac{5}{x^2+4}\right)^n$ converges?

(A) $x < -\frac{1}{2}$ and $x > \frac{1}{2}$

(B) $x < -1$ and $x > 1$

(C) $-\frac{1}{2} < x < \frac{1}{2}$

(D) $-1 < x < 1$

END OF PART A OF SECTION I

STOP

CALCULUS BC TEST 6
SECTION I, Part B
Time — 45 minutes
Number of questions — 15

A GRAPHING CALCULATOR IS REQUIRED FOR SOME QUESTIONS ON THIS PART OF THE EXAM.

Directions: Solve each of the following problems using the available space for scratch work. Choose the best answer among the answer choices given and fill in the corresponding circle on the answer sheet.

76. If f is a twice differentiable function such that $f'(1) = 3$ and $f''(1) = 3$, all of the following are true EXCEPT

(A) $\lim\limits_{h \to 0} \dfrac{f(1+h) - f(1)}{h} = 3$

(B) $\lim\limits_{x \to 1} \dfrac{f(x) - f(1)}{x - 1} = 3$

(C) $\lim\limits_{h \to 0} \dfrac{f'(1+h) - f'(1)}{h} = 3$

(D) $\lim\limits_{x \to 1} f(x) = f(1) = 3$

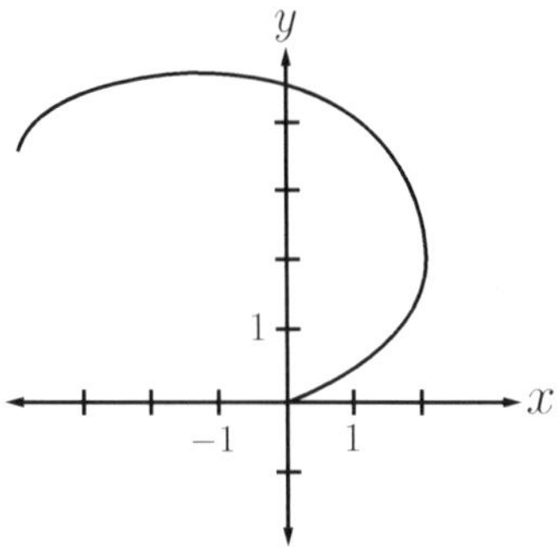

77. The figure above shows a portion of the polar curve $r = \theta + 3\sin\theta$. Which of the following is the area of the region bounded by the curve and the y-axis ?

(A) 5.302 (B) 6.103 (C) 6.497 (D) 7.180

---

78. If $\int_{10}^{2} f(x)\,dx = -5$ and $\int_{7}^{10} f(x)\,dx = -3$, then $\int_{2}^{7} 2f(x) - 4\,dx =$

(A) 12 (B) 8 (C) −4 (D) −6

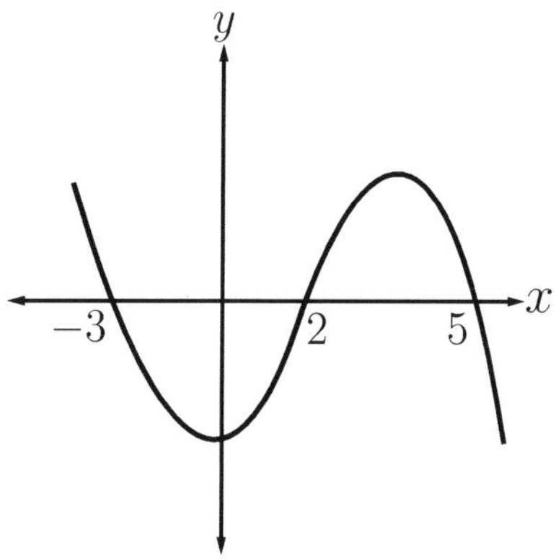

79. The graph of f', the derivative of f, is shown above. If the derivative of h, h', is defined by $h'(x) = f'(1-x)$, for which of the following values of x does h attain a relative minimum?

(A) $x = 2$ only

(B) $x = 1$ only

(C) $x = -2$ and $x = 6$

(D) $x = -4$ and $x = 4$

---

80. At time $t > 0$, a particle moving in the xy-plane has vector velocity given by $< \dfrac{t-1}{t}, \sqrt{t^3+t} >$. Which of the following is the acceleration of the particle at time $t = 2$?

(A) $< 0.25, 1.716 >$

(B) $< 0.25, 2.055 >$

(C) $< 0.25, 2.316 >$

(D) $< 0.25, 2.576 >$

81. The growth rate, $\frac{dP}{dt}$, of a population of meerkats in a desert is modeled by a logistic differential equation. At $t = 3$ years, the number of meerkats is 200 and is increasing at the rate of 200 meerkats per year. If the maximum population of meerkats in the desert is 800, which of the following differential equation best describes the population of meerkats?

(A) $\frac{dP}{dt} = \frac{1}{600}P(P - 400)$

(B) $\frac{dP}{dt} = \frac{1}{400}P(P - 400)$

(C) $\frac{dP}{dt} = \frac{1}{600}P(800 - P)$

(D) $\frac{dP}{dt} = \frac{1}{400}P(800 - P)$

---

82. If $f(x) = (\ln x)^2$, then $f''\left(\frac{1}{e}\right) =$

(A) $4e^2$ (B) $2e^2$ (C) $\frac{1}{2\sqrt{e}}$ (D) $\frac{1}{4\sqrt{e}}$

x	$x<1$	1	$1<x<2$	2	$2<x<3$	3	$x>3$
$f(x)$	Positive	9	Positive	5	Positive	1	Positive
$f'(x)$	Positive	0	Negative	Negative	Negative	0	Positive
$f''(x)$	Negative	Negative	Negative	0	Positive	Positive	Positive

83. Let f be a twice differentiable function. The table above shows values of f and its derivatives, f' and f'', for selected values of x. Which of the following statement must be true?

(A) There must be c in the interval $(2, 3)$ such that $f'(c) = 0$.

(B) f' is decreasing on the interval $(1, 3)$.

(C) f has a local maximum at the point $(3, 1)$

(D) f is decreasing and concave up on the interval $(2, 3)$.

---

84. Which of the following is the area of the region enclosed by the graphs $y = 2x^3 - 10x^2 + 4x + 16$ and $y = e^x$?

(A) 20.707 (B) 21.393 (C) 23.426 (D) 25.927

85. Let f be a positive, continuous, and decreasing function on the closed interval $[0, 2]$. If the function g is defined by $g(x) = \int_1^x f(t)\,dt$, which of the following table best represents the values of g ?

(A)

x	$g(x)$
0	-3
1	0
2	5

(B)

x	$g(x)$
0	-3
1	0
2	1

(C)

x	$g(x)$
0	3
1	0
2	1

(D)

x	$g(x)$
0	3
1	0
2	5

---

86. The rate at which people enter the Natural history museum is given by $R(t) = 5\left(t + \frac{1}{t}\right)^2$ for $t \geq 1$, where t is measured in minutes and $R(t)$ is measured in the number of people per minute. If none of people leave the museum, which of the following best approximates the average number of people entering the museum from $t = 1$ minute to $t = 11$ minutes?

(A) 196 (B) 218 (C) 232 (D) 257

87. Let f be a function that has derivative of all orders such that $|f^{(n)}(x)| < 5$ for all real numbers x. Suppose $R_n(x)$ be the error defined by $|f(x) - T_n(x)|$, where T_n is the nth degree Taylor polynomial for f about $x = 2$. What is the maximum error $R_3(1.8)$ using the Lagrange error bound?

(A) $\frac{1}{5^2 \cdot 3!}$ (B) $\frac{1}{5 \cdot 4!}$ (C) $\frac{1}{5^2 \cdot 4!}$ (D) $\frac{1}{5^3 \cdot 4!}$

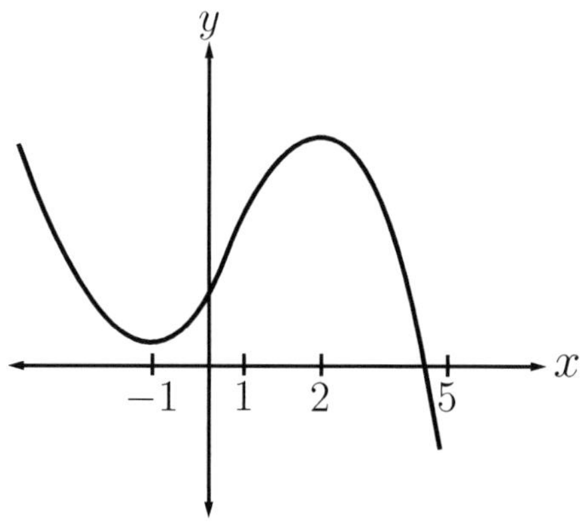

Graph of f

88. The figure above shows the graph of a function f. At which of the following value of x does f satisfy $f''(x) < f'(x) < f(x)$?

(A) -1 (B) 1 (C) 2 (D) 3

89. If the function f is defined by $f(x) = x^2 \sin(2x)$, which of the following is the Taylor series for f about $x = 0$?

(A) $2x^3 - \dfrac{8x^5}{3!} + \dfrac{32x^7}{5!} - \dfrac{128x^9}{7!} + \cdots$

(B) $2x^3 - \dfrac{4x^5}{3!} + \dfrac{16x^6}{5!} - \dfrac{64x^9}{7!} + \cdots$

(C) $x^2 - \dfrac{4x^4}{2!} + \dfrac{16x^6}{4!} - \dfrac{64x^8}{6!} + \cdots$

(D) $x^2 - \dfrac{2x^4}{2!} + \dfrac{8x^6}{4!} - \dfrac{32x^8}{6!} + \cdots$

---

90. If a curve defined by the parametric equations $y(t) = f(t)$ and $x(t) = e^{-t}$, then $\dfrac{d^2y}{dx^2} =$

(A) $\dfrac{f''(t) - tf'(t)}{e^t}$

(B) $\dfrac{f''(t) + tf'(t)}{e^{2t}}$

(C) $e^t\big(f''(t) - f'(t)\big)$

(D) $e^{2t}\big(f''(t) + f'(t)\big)$

END OF PART B OF SECTION I

STOP

CALCULUS BC TEST 7
SECTION I, Part A
Time — 60 minutes
Number of questions — 30

A CALCULATOR MAY NOT BE USED ON THIS PART OF THE EXAM.

Directions: Solve each of the following problems using the available space for scratch work. Choose the best answer among the answer choices given and fill in the corresponding circle on the answer sheet.

1. Which of the following is the particular solution to the equation $f'(x) = \dfrac{x+1}{x}$ with the initial condition $f(1) = 3$?

(A) $x + 2$

(B) $x + \ln|x| + 2$

(C) $e^{x-1} + 3$

(D) $\sqrt{x-1} + 3$

2. For $x > 2$, which of the following is the x-coordinate of the point on the graph $y = \dfrac{1}{x-2}$ where the tangent line is parallel to the secant line that passes through the graph at $x = 3$ and $x = 4$?

(A) $2+\sqrt{3}$ (B) $2+\sqrt{2}$ (C) $2-\sqrt{2}$ (D) $2-\sqrt{3}$

---

3. If $f(x) = \sin^{-1}\left(\dfrac{x}{3}\right)$, then $f'(x) =$

(A) $\dfrac{1}{\sqrt{9-x^2}}$ (B) $\dfrac{-3}{\sqrt{9-x^2}}$ (C) $\dfrac{1}{\sqrt{x^2-9}}$ (D) $\dfrac{-1}{3\sqrt{x^2-9}}$

4. Which of the following gives the area of the region in the second quadrant enclosed by the polar curve $r = \sin\theta$, the y-axis, and the line $y = -x$?

(A) $\displaystyle\int_{\frac{\pi}{4}}^{\frac{\pi}{2}} \sqrt{\sin^2\theta + \cos^2\theta}\, d\theta$

(B) $\displaystyle\int_{\frac{\pi}{2}}^{\frac{3\pi}{4}} \sqrt{\sin^2\theta + \cos^2\theta}\, d\theta$

(C) $\displaystyle\frac{1}{2}\int_{\frac{\pi}{4}}^{\frac{\pi}{2}} \sin^2\theta\, d\theta$

(D) $\displaystyle\frac{1}{2}\int_{\frac{\pi}{2}}^{\frac{3\pi}{4}} \sin^2\theta\, d\theta$

---

5. Let f be a function with $f(2) = 1$, $f'(2) = -4$, $f''(2) = 8$, and $f'''(2) = -24$. Which of the following is the third degree Taylor polynomial for f about $x = 2$?

(A) $1 - 4(x-2) + 4(x-2)^2$

(B) $1 - 4(x-2) + 8(x-2)^2$

(C) $1 - 4(x-2) + 4(x-2)^2 - 4(x-2)^3$

(D) $1 - 4(x-2) + 8(x-2)^2 - 24(x-2)^3$

6. Consider the series $\sum_{n=1}^{\infty} \frac{4^n}{n!}$. If the ratio test is applied to determine whether the series converges, which of the following inequality results?

(A) $\lim_{n\to\infty} \frac{4}{(n+1)!} < 1$

(B) $\lim_{n\to\infty} \frac{4}{n+1} < 1$

(C) $\lim_{n\to\infty} \frac{n+1}{4} < 1$

(D) $\lim_{n\to\infty} 4(n+1) < 1$

---

7. $\lim_{x\to 0} \frac{4\sin x \cos x}{3x} =$

(A) $\frac{4}{3}$ (B) 1 (C) $\frac{2}{3}$ (D) $\frac{1}{2}$

8. A particle moves in the x-axis so that the velocity of the particle is given by $v(t) = (t+1)^2(t-3)$. How many times does the particle change its direction from $t = -5$ to $t = 15$?

(A) 3 (B) 2 (C) 1 (D) 0

---

9. $$\int x^2 e^x \, dx =$$

(A) $e^x(x-1)^2 + C$

(B) $e^x(x^2 - 2x + 2) + C$

(C) $e^x(x^2 - x + 1) + C$

(D) $e^x(x^2 - x - 1) + C$

10. Which of the following is the value of p so that the series $\displaystyle\sum_{n=1}^{\infty} \frac{n^3+1}{n^p+3}$ converges?

(A) $p \geq 3$ (B) $p > 3$ (C) $p \geq 4$ (D) $p > 4$

---

11. A population of deer in a forest grows according to the differential equation $\dfrac{dP}{dt} = 0.05P(500-P)$, where P is the number of deer at time t in years. Which of the following statements are true?

I. The maximum number of deers in the forest is 500.

II. The population of deer grows fastest when the number of deer is 250.

III. The number of deer in the forest at $t = 0$ year is 100.

(A) II only

(B) III only

(C) I and II only

(D) II only III only

12. Let f be a twice differentiable function with $f''(x) = \sqrt{x^2+1}(x^2-5x)$ and have critical numbers at $x=1$, $x=3$, and $x=6$. At which of the following values of x does f have a local maximum?

(A) $x=3$ and $x=6$

(B) $x=1$ and $x=3$

(C) $x=6$ only

(D) $x=1$ only

---

13. For any time $t \geq 0$, water is pumped into a tank at the rate $R(t) = 5(t+1)^4$ cubic feet per minute, where t is measured in minutes. If the tank has 275 cubic feet at $t=1$ minute, how much water does the tank have in the beginning?

(A) 244 (B) 212 (C) 186 (D) 141

14. Let f and g be continuous functions for all real numbers x. If $\int_b^c f(x)\,dx = 5$, $\int_b^a f(x)\,dx = -3$, $\int_b^c g(x)\,dx = 6$, and $\int_a^b g(b)\,dx = -1$, then $\int_a^c (f(x) - 2g(x))\,dx =$

(A) -4 (B) -2 (C) 1 (D) 3

---

x	f	g	f'	g'
-1	2	1	-1	3
1	1	3	0	-2
2	-2	-1	2	1

15. The table above shows values of the differentiable functions f and g and of their derivatives f' and g' for selected values of x. If the function h is defined by $h(x) = f(g(2x))$, which of the following is the slope of the tangent line to the graph of h at $x = 1$?

(A) 4 (B) 3 (C) -2 (D) -5

16. The region R is enclosed by the graph of $y = \dfrac{1}{2\sqrt[4]{x}}$, the x-axis, and the lines $x = 1$ and $x = k$, where $k > 1$. If the volume of the solid obtained by rotating the region R about the x-axis is π, which of the following is the value of k ?

(A) 2 (B) 4 (C) 8 (D) 9

---

17. Let f be a differentiable function for all real numbers x. If $\lim_{x \to 3} \dfrac{x^2 f(x) - 9}{x - 3} = 3$, which of the following is the value of $f'(3)$?

(A) 3 (B) 2 (C) $\dfrac{1}{2}$ (D) $-\dfrac{1}{3}$

18. Which of the following sequence converges?

I. $\left\{\dfrac{n^2+3n+1}{3n^2-n+1}\right\}_1^\infty$ II. $\left\{\dfrac{n}{\ln n}\right\}_1^\infty$ III. $\left\{\left(1+\frac{1}{n}\right)^n\right\}_1^\infty$

(A) I only

(B) II only

(C) I and III only

(D) II and III only

---

19. If $f(x) = \dfrac{x^2\sqrt{x+1}}{3x-1}$, then $f'(x) =$

(A) $\dfrac{x^2\sqrt{x+1}}{3x-1}\left(2x - \dfrac{2}{(x+1)} + \dfrac{1}{(3x-1)}\right)$

(B) $\dfrac{x^2\sqrt{x+1}}{3x-1}\left(2x + \dfrac{2}{(x+1)} + \dfrac{3}{(3x-1)}\right)$

(C) $\dfrac{x^2\sqrt{x+1}}{3x-1}\left(\dfrac{2}{x} + \dfrac{1}{2(x+1)} - \dfrac{3}{3x-1}\right)$

(D) $\dfrac{x^2\sqrt{x+1}}{3x-1}\left(\dfrac{2}{x} + \dfrac{1}{2(x+1)} + \dfrac{1}{3(3x-1)}\right)$

20. If $x^3 + y^3 = 9$, which of the following is the value of $\dfrac{d^2y}{dx^2}$ at the point $(1, 2)$?

(A) $-\dfrac{9}{16}$ (B) $-\dfrac{9}{32}$ (C) $\dfrac{5}{32}$ (D) $\dfrac{3}{16}$

---

21. All of the following statements are true EXCEPT

(A) The series $\displaystyle\sum_{n=1}^{\infty}(-1)^n \frac{3}{\sqrt[3]{n^2+2}}$ is absolutely convergent.

(B) The series $\displaystyle\sum_{n=1}^{\infty}(-1)^n \frac{2}{n+3}$ is conditionally convergent.

(C) The series $\displaystyle\sum_{n=1}^{\infty} \frac{2^{n+1}}{3^{n-2}}$ is convergent.

(D) The series $\displaystyle\sum_{n=1}^{\infty} \left(\frac{n+1}{n}\right)^n$ is divergent.

22. Which of the following is the radius of convergence of the Maclaurin series for $f(x) = \dfrac{1}{1-x}$?

(A) 0 (B) $\dfrac{1}{2}$ (C) 1 (D) ∞

---

23. Which of the following function has the Taylor series $\dfrac{x^3}{2!} - \dfrac{x^5}{4!} + \dfrac{x^7}{6!} + \cdots + (-1)^{n+1}\dfrac{x^{2n+1}}{(2n)!} + \cdots$ about $x = 0$?

(A) $xe^x - 1$

(B) $1 - e^{-x}$

(C) $x(1 - \sin x)$

(D) $x(1 - \cos x)$

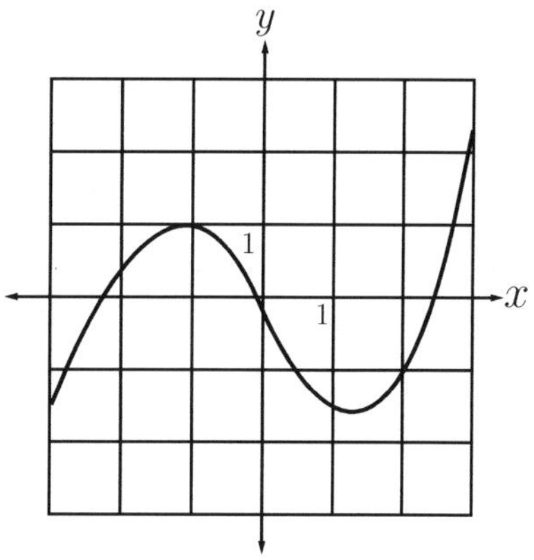

Graph of f

24. The figure above shows the graph of the function f for $-3 \le x \le 3$. If the function g is defined by $g(x) = \int_1^{2x} f(t)\,dt$, which of the following is the value of $g'(1)$?

(A) 2 (B) -1 (C) -2 (D) -3

---

25. A curve is defined by the parametric equations $x = t^2 - 2t + 3$ and $y = \sqrt{t-1}$. Which of the following is the equation of the line tangent to the curve at the point $(18, 2)$?

(A) $y - 2 = \frac{1}{64}(x - 18)$

(B) $y - 2 = \frac{1}{32}(x - 18)$

(C) $y - 2 = -\frac{3}{16}(x - 18)$

(D) $y - 2 = -\frac{3}{8}(x - 18)$

26. Which of the following is a power series expansion for $\dfrac{1}{(1-x)^2}$?

(A) $1 - 2x + 3x^2 - 4x^3 + \cdots + (-1)^{n+1}nx^{n-1} + \cdots$

(B) $1 + 2x + 3x^2 + 4x^3 + \cdots + nx^{n-1} + \cdots$

(C) $1 - x^2 + x^4 - x^6 + \cdots + (-1)^{n+1}x^{2n-2} + \cdots$

(D) $1 + x^2 + x^4 + x^6 + \cdots + x^{2n-2} + \cdots$

---

27. Which of the following is the average value of $f(x) = x^2\sqrt{x^3 - 1}$ on the interval $[1, 3]$?

(A) $\dfrac{3}{2}\sqrt{3}$ (B) $\dfrac{11}{5}\sqrt{11}$ (C) $\dfrac{13}{8}\sqrt{13}$ (D) $\dfrac{26}{9}\sqrt{26}$

28. If the function f is given by $f(x) = \frac{1}{4}x^4 - \frac{3}{2}x^2 + 1$, on which of the following intervals is f increasing?

(A) $(-\infty, -\sqrt{3})$ only

(B) $(-\sqrt{3}, 0)$ only

(C) $(-\infty, -\sqrt{3})$ and $(0, \sqrt{3})$

(D) $(-\sqrt{3}, 0)$ and $(\sqrt{3}, \infty)$

---

29. Let f be the function given by $f(x) = \sqrt[3]{x-1} + k$. If $\int_0^2 f(x)\,dx = 14$, which of the following is the value of k ?

(A) 3 (B) 4 (C) 6 (D) 7

x	1	2	3	4
$g(x)$	2	k	4	9

30. Let g be a continuous function on the closed interval $[1, 4]$. The values of g for selected values of x are shown in the table above. Which of the following is the value of k so that the equation $g(x) = 5$ has at least three solutions on the interval $[1, 4]$?

(A) 3 (B) 4 (C) 5 (D) 6

END OF PART A OF SECTION I

STOP

CALCULUS BC TEST 7
SECTION I, Part B
Time — 45 minutes
Number of questions — 15

A GRAPHING CALCULATOR IS REQUIRED FOR SOME QUESTIONS ON THIS PART OF THE EXAM.

Directions: Solve each of the following problems using the available space for scratch work. Choose the best answer among the answer choices given and fill in the corresponding circle on the answer sheet.

76. The rate at which the circumference of a circle is increasing is 4 cm/sec. How fast is the area of the circle increasing when the circumference of the circle is 6π ?

(A) 12 cm^2/sec

(B) 9 cm^2/sec

(C) 8 cm^2/sec

(D) 6 cm^2/sec

x	1	2	3	4
$f(x)$	5	-1	2	3

77. The values of f at selected values of x are shown in the table above. If the function g is defined by $g(x) = \int_{3x}^{x^2} f(t)\,dt$, which of the following is the value of $g'(1)$?

(A) -3 (B) -1 (C) 4 (D) 5

78. Let f be the function defined by $f(x) = \sqrt{x+1}$. Which of the following is the value of c such that the instantaneous rate of change of f equals the average rate of change of f over the interval $[0, 3]$?

(A) 0.37 (B) 0.75 (C) 0.94 (D) 1.25

79. If the first derivative of f, f', is given by $f'(x) = \dfrac{\sin 2x}{x} - \dfrac{1}{2}$, how many critical numbers does f' have on the closed interval $[-5, 5]$?

(A) 4 (B) 3 (C) 2 (D) 1

---

80. An electric company manufactured a tablet computer at the rate of $\dfrac{e^t + 3}{\sqrt[3]{t^2 + 1}}$ tablet computers per day, where t is measured in days. To the nearest integer, how many tablet computers did the electric company manufacture during the time interval $1 \le t \le 9$?

(A) 2425 (B) 2046 (C) 1749 (D) 1510

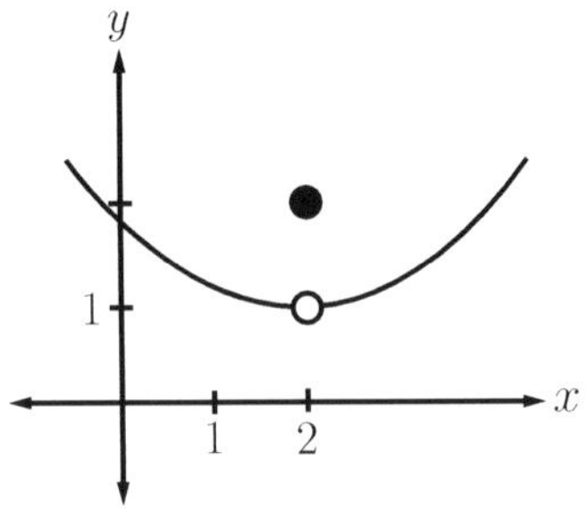

Graph of g

81. If the graph of the function g is shown above, what is the value of $\lim_{x \to 2} \cos\left(g(x)\right)$?

(A) 0.841 (B) 0.540 (C) -0.384 (D) Nonexistent

---

82. A particle moves in the xy-plane so that its position at any time t is given by $x = \sin(t^2)$ and $y = e^{-2t}$. Which of the following is the speed of the particle at $t = 3$?

(A) 5.467 (B) 5.698 (C) 6.351 (D) 6.927

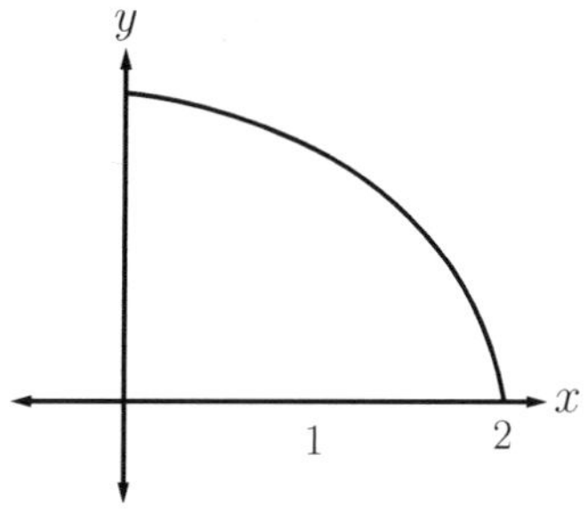

Graph of f

83. The graph of the function f is shown above. Which of the following must be true about the approximation of $\int_0^2 f(x)\,dx$?

(A) The trapezoidal sum is an overestimate and a left Riemman sum is an overestimate.

(B) The trapezoidal sum is an overestimate and a left Riemman sum is an underestimate.

(C) The trapezoidal sum is an underestimate and a left Riemman sum is an underestimate.

(D) The trapezoidal sum is an underestimate and a left Riemman sum is an overestimate.

---

84. A particle moves along the x-axis so that its velocity is given by $v(t) = \sin(t^3 - 2)$ for any time $t \geq 0$. If the position of the particle at $t = 0$ is -2, which of the following is the position of the particle when its velocity is equal to 0 for the first time?

(A) -3.453 (B) -3.072 (C) 0.928 (D) 1.489

85. The base of a solid is the region bounded by the graphs of $y = x(4 - x)$ and $y = x$. Which of the following is the volume of the solid if cross-sections perpendicular to the x-axis are equilateral triangles?

(A) 3.507 (B) 3.778 (C) 4.305 (D) 4.711

---

86. Let f be the function defined by $f(x) = \sqrt{e^x + 3}$. Which of the following best approximates the value of $f(1.25)$ using the line tangent to the graph of f at $x = 0$?

(A) 2.059 (B) 2.107 (C) 2.270 (D) 2.313

87. Which of the following is the interval of convergence of the series $\displaystyle\sum_{n=1}^{\infty} \frac{3^n(x-1)^n}{n^3}$?

(A) $\dfrac{2}{3} \le x < \dfrac{4}{3}$

(B) $\dfrac{2}{3} \le x \le \dfrac{4}{3}$

(C) $-2 < x \le 4$

(D) $-2 \le x \le 4$

---

88. If $\displaystyle\int_1^3 x^n \, dx = k$, which of the following is the value of $\displaystyle\int_1^2 (5-2x)^n \, dx$ in terms of k ?

(A) $-\dfrac{k}{2}$ (B) $\dfrac{k}{2}$ (C) k (D) $\dfrac{3k}{2}$

89. Which of the following is the value of $\lim_{x\to\infty} \dfrac{x^{10}}{e^x}$?

(A) 0 (B) e (C) 10! (D) nonexistent

90. Which of the following is the second degree Taylor polynomial for $f(x) = \ln x$ about $x = 2$?

(A) $\ln 2 + \frac{1}{2}(x-2) - \frac{1}{8}(x-2)^2$

(B) $\ln 2 + \frac{1}{2}(x-2) + \frac{1}{8}(x-2)^2$

(C) $\ln 2 - \frac{1}{2}(x-2) - \frac{1}{4}(x-2)^2$

(D) $\ln 2 - \frac{1}{2}(x-2) + \frac{1}{4}(x-2)^2$

END OF PART B OF SECTION I

STOP

Answers

AP CALCULUS BC Test 1

Answers

1. A	11. A	21. D	76. C	86. A
2. C	12. A	22. D	77. D	87. B
3. A	13. D	23. C	78. C	88. D
4. A	14. B	24. A	79. D	89. A
5. B	15. A	25. B	80. A	90. C
6. D	16. D	26. A	81. B	
7. D	17. A	27. D	82. C	
8. C	18. C	28. B	83. A	
9. B	19. C	29. C	84. B	
10. D	20. B	30. D	85. B	

Answers

AP CALCULUS BC Test 2

Answers

1. C	11. B	21. D	76. D	86. C
2. B	12. D	22. B	77. C	87. B
3. B	13. D	23. C	78. A	88. B
4. C	14. B	24. B	79. B	89. C
5. A	15. C	25. D	80. B	90. D
6. A	16. B	26. A	81. B	
7. C	17. C	27. C	82. A	
8. A	18. A	28. C	83. D	
9. B	19. D	29. B	84. A	
10. C	20. B	30. D	85. D	

Answers

AP CALCULUS BC Test 3

Answers

1. B	11. B	21. B	76. A	86. A
2. D	12. D	22. C	77. D	87. C
3. C	13. B	23. B	78. B	88. D
4. B	14. D	24. A	79. C	89. B
5. A	15. D	25. B	80. C	90. A
6. D	16. C	26. C	81. D	
7. D	17. D	27. C	82. A	
8. C	18. A	28. B	83. A	
9. A	19. C	29. B	84. C	
10. D	20. D	30. D	85. B	

Answers

AP CALCULUS BC Test 4

Answers

1. D	11. B	21. D	76. C	86. C
2. A	12. D	22. C	77. A	87. D
3. D	13. D	23. C	78. C	88. A
4. B	14. B	24. D	79. D	89. B
5. D	15. A	25. B	80. B	90. D
6. A	16. B	26. C	81. B	
7. A	17. B	27. A	82. D	
8. A	18. A	28. D	83. C	
9. D	19. C	29. B	84. B	
10. B	20. C	30. A	85. A	

Answers

AP CALCULUS BC Test 5

Answers

1. D	11. B	21. D	76. C	86. C
2. C	12. D	22. D	77. C	87. A
3. C	13. B	23. A	78. B	88. C
4. A	14. C	24. A	79. D	89. C
5. A	15. D	25. B	80. B	90. D
6. C	16. B	26. D	81. C	
7. D	17. D	27. A	82. A	
8. D	18. A	28. A	83. D	
9. B	19. A	29. C	84. A	
10. B	20. C	30. D	85. D	

Answers

AP CALCULUS BC Test 6

Answers

1. D	11. B	21. D	76. D	86. C
2. D	12. B	22. C	77. D	87. D
3. D	13. A	23. C	78. C	88. C
4. B	14. D	24. B	79. D	89. A
5. B	15. C	25. D	80. B	90. D
6. A	16. B	26. A	81. C	
7. C	17. B	27. C	82. A	
8. C	18. A	28. A	83. D	
9. B	19. C	29. D	84. D	
10. D	20. C	30. B	85. B	

Answers

AP CALCULUS BC Test 7

Answers

1. B	11. C	21. A	76. A	86. D
2. B	12. B	22. C	77. C	87. B
3. A	13. A	23. D	78. D	88. B
4. D	14. B	24. C	79. C	89. A
5. C	15. C	25. B	80. B	90. A
6. B	16. D	26. B	81. B	
7. A	17. D	27. D	82. A	
8. C	18. C	28. D	83. D	
9. B	19. C	29. D	84. B	
10. D	20. A	30. D	85. A	

74852993R00104

Made in the USA
San Bernardino, CA
21 April 2018